The
Weather
Factor

Books by David M. Ludlum

The Weather Factor
New Jersey Weather
The American Weather Book
The Country Journal New England Weather Book
Weather Record Book
Early American Tornadoes
Early American Winters I and II
Early American Hurricanes
Social Ferment in Vermont

The
Weather
Factor

David M. Ludlum

Houghton Mifflin Company   Boston   1984

Printed in the United States of America

V 10 9 8 7 6 5 4 3 2 1

Library of Congress Cataloging in Publication Data

Ludlum, David McWilliams, date
  The weather factor.

  1. United States—Climate—History. I. Title.
QC983.L83   1984   551.6973'09   83-26658
ISBN 0-395-27604-7
ISBN 0-395-36144-3 (pbk.)

Quotations on pp. 231–238, from Charles Lindbergh, *The Spirit of St. Louis.* Copyright 1953 Charles Scribner's Sons; copyright renewed 1981 by Anne Morrow Lindbergh. Reprinted with the permission of Charles Scribner's Sons.

The passage on pp. 230–231 is reprinted by permission of McGraw-Hill Book Company from *The Papers of Robert H. Goddard* by Robert H. Goddard (New York: McGraw-Hill, 1970).

The passage on pp. 248–249 is reprinted by permission of Dodd Mead & Co. from *Up Ship* by C. E. Rosendahl (New York: Dodd Mead, 1931).

The passage on pp. 255–256 is reprinted by permission of the U.S. Naval Institute from *The Airships Akron and Macon* by Richard K. Smith. Copyright © 1965, U.S. Naval Institute, Annapolis, Maryland.

The passage on pp. 266–267 is reprinted by permission of Wadsworth Publishing Co. from *The Saga of the Air Mail* by Carroll V. Glines (Princeton: D. Van Nostrand, 1968).

The Revolutionary War section contains excerpts from articles previously published in *Weatherwise* and reprinted here by permission of Weatherwise, Inc., of Princeton, N.J.

Contents

CHAPTER FIVE

The Realm of Flight · 208

A hard frost, a sudden thaw, a "hot spell," a "cold snap," a contrary wind, a long drought, a storm of sand, — all these things have had their part in deciding the destinies of dynasties, the fortunes of races, the fate of nations. Leave the weather out of history, and it is as if night were left out of day, and winter out of the year.

— Charles C. Hazewell, *Atlantic Monthly*, May 1862

The Weather Factor

CHAPTER ONE

The Colonial Period

First Encounters with the American Climate

When the sun rose out of the broad expanse of the North Atlantic Ocean on the first day of the new century in January 1601, its rays fell on a vast continent whose geography and climate were unknown to Europeans. Though the waters of the coastline had been explored in summertime, and fishermen visited the Grand Banks of Newfoundland annually, no white men were wintering on the shores of the northwestern reaches of the Atlantic Ocean, nor had any previous visitor left any historical record of the extremities of the seasons, aside from a few words of Jacques Cartier about his stayover at Quebec in the winter of 1535–36.

Of the territory to be the future United States, only single settlements of the Spanish in Florida and New Mexico existed, and these were but isolated outposts close to the tropics, where benign weather conditions generally prevailed. The nature of atmospheric behavior over the American continent would remain a great unknown until the descendants of the first settlers, in crossing the mountains and traversing the waterways of the interior during the next two centuries, finally pieced together bits of climate information that eventually brought knowledge of the whole into focus.

New France

The French were the first to make a permanent settlement on the northern Atlantic Coast, choosing in 1604 an island in the St. Croix River between the present State of Maine and the Province of New Brunswick. The establishment of two colonies by the English soon followed in 1607, one of temporary duration in Maine and the other destined to be permanent in Virginia. The French were forced to remove from their original site, as were the British from Maine, as a result of their harsh experience with a severe continental winter. The French resettled in the milder, maritime exposure of Annapolis in Nova Scotia, but the English abandoned their colony and returned home.

The testimony of Samuel de Champlain, who served as geographer of the first French expedition, told much about the climate: "It was difficult to know this country without having wintered there; for on arriving in summer everything is very pleasant on account of the woods, the beautiful landscapes, and the fine fishing for the many kinds of fish found there. There are six months of winter in that country." He further declared: "The cold was severe and more extreme than in France, and lasted much longer."

As a result of the winter experience of the French expedition, Champlain sailed as far south as Cape Cod the next summer to explore the area and question the natives about the nature of the climate. He concluded: "This region is of moderate temperature and the winter not severe," but of the Penobscot Bay area of Maine, he wrote: "I believe this region is as disagreeable in winter as is that of our settlement, in regard to which we were greatly deceived." Sieur de Monts, commander of the expedition, in deciding to "build another settlement to escape the cold and dreadful winter we experienced at Ste. Croix island . . . searched for a suitable site for our residence, with shelter from the northwest wind, which we dreaded on account of our having been greatly distressed thereby." The settlement at Annapolis proved permanent.

The first person to attempt a description of the general outlines of the New England climate was Captain John Smith, the famed soldier and romantic figure of the Virginia Colony, who explored the Massachusetts and Maine coastlines in 1614. He gathered much valuable information about the face of the country and its resources, which he published in *A Description of New England* in 1616. Smith described the bountiful fisheries, fertile soil, and varied products under the heading "A Proof of an Excellent Climate." He was especially rapturous about what he observed in the vicinity of Massachusetts Bay. In noting "the moderate temper of the air," he asked, "Who can but approve this a most excellent place, both for health and fertility? And of the four parts of the world that I have yet seen not inhabited, could I but transport a colony, I would rather live here than any where."

Another pre-Plymouth appraisal of the climate was contained in *A Brief Relation of the Discovery and Plantation of New England*, a tract intended for real estate promotion of the Maine region. Its anonymous author stated: "But this country, what by the general and particular situation, is so temperate, as it seemeth to hold the golden mean." The health and happiness of those settling there "can be imputed to no other cause, than the temperature of the climate," he concluded.

Maine

The first group of Englishmen known to have wintered in New England established the Sagadahoc Colony on the central Maine coast near the mouth of the Kennebec River in August 1607. The site is now marked by the restored Fort Popham and Popham Beach, named for the leader of the expedition, Sir George Popham. Unfortunately, he did not survive the rigors of the very severe winter ensuing, so command of the expedition passed to Captain Raleigh Gilbert. A brief reference to the unfavorable weather conditions experienced by the colony was contained in William Strachey's *Historie of Travaile into Virginia Britannia*, published in London, probably in 1612: ". . . had not the winter proved so extreme unseasonable and frosty; for it was the year 1607, when the extraordinary frost was felt in most parts of Europe, it was here likewise as vehement, by which no boat could stir upon any business."

In the succeeding summer Gilbert heard of the death of his brother and felt the call to return home to take up his inheritance. And because of their "fear that all the other winters would prove like the first, the company would by no means stay in the country," and the entire colony embarked for England. Sagadahoc fell victim to the North American winter; its name might have taken symbolic precedence in New England over Plymouth had the winter of 1607–08 been of a less severe nature.

Plymouth and Massachusetts Bay

The question of the general climate and degree of severity of winter weather remained uppermost in the minds of planners and planters of new colonies. The success of their undertakings hung in the balance, much depending on the ability of the settlers to cope with the meteorological aspect of the American scene.

The advanced date of the season in November 1620 proved most influential in the minds of the Pilgrim Fathers when they were making the fateful decision not to continue on to their intended destination within the lands of the Virginia Company, but to remain on Cape Cod and seek a settlement site in the vicinity. In describing the predicament faced by the Mayflower band under the gray skies of November, Governor William Bradford composed one of the most moving passages in all American history in a "back-to-the-wall" journal entry:

> Being thus passed the vast ocean . . . they had now no friends to welcome them nor inns to entertain or refresh their weatherbeaten bodies; no houses or much less towns to repair to, to seek for succor . . . And for the season it was winter, and they that know the winters of that country know them to be sharp and violent, and subject to cruel and fierce storms, dangerous to travel to known places, and much more to search an unknown coast.

Though the early New Englanders were prolific in producing documents and commentaries on the political and theological affairs of their colonies, little was written about the weather and climate that has value today. Of course, the lack of instruments and the undeveloped state of all science in that age militated against the production of worthwhile treatises. Two writers did attempt the subject: William Wood in *New England Prospects*, published in London in 1634, and John Josselyn in *An Account of Two Voyages to New-England made during the years 1638 and 1663*.

Chapter Two of Wood's *Prospects* was devoted to: "Of the seasons of the year, Winter and Summer, together with the Heat, Cold, Snow, Rain, and the effects of it." He took a most favorable, almost propagandistic, view of the region's climate, declaring:

> For that part of the country wherein most of the English have their habitations: it is for certain the best ground and sweetest climate in all those parts, bearing the name of New England, agreeing well with the temper of our English bodies, being high land, and sharp air, and though most of our English towns border upon the sea-coast, yet they are not often troubled with mists, or unwholesome fogs, or cold weather from the sea, which lies east and south from the land . . . only the northwest wind coming from the land, is the cause of extreme cold weather, being always accompanied with deep snows and bitter frost, so that in two or three days the rivers are passable for horse and man. But it is an axiom in Nature, *Nullum violentum est perpetuum* (No extremes last long), after which the weather is more tolerable, the air being nothing so sharp; but peradventure in four or five days after those cold messengers will blow afresh, commanding every man to his house, forbidding any to outface him without prejudice to their noses.

Wood replies to those who claim the country may be too cold for Englishmen by pointing out "there is a good store of wood with which to make fires to warm and houses to live in." He says also, "the cold lasts only two months, and every ten years there is a very mild season" such as favored the Pilgrims in 1620–21 and the Puritans in 1629–30. He emphasizes that the cold in America does not produce "so many noisome effects as the raw winter of England. In public assemblies it is strange to hear a man sneeze or cough as ordinarily they do in England."

Wood also had some words to say about the American summer, which was a new experience for most colonists from northwestern Europe:

> The summers be hotter than in England, because of their more southern latitude, yet they are tolerable, being often cooled with fresh blowing winds, it seldom being so hot as men are driven from

their labors, especially such whose employments are within doors or under the cool shade. . . . The summers are commonly hot and dry, there being seldom any rains; I have known it six or seven weeks before one shower hath moistened the plowman's labor, yet the harvest hath been very good, the Indian corn requiring more heat than wet. For the English corn [wheat], it is refreshed with the nightly dews till it grow up to shade his roots with his own substance from the parching sun.

John Josselyn, or "Old Josselyn" as he became affectionately known later, was the author of two books dealing with New England based on his observations during two visits, in 1638–39 and 1663–71. His *Account of Two Voyages to New-England* contained a rather strange compound of scientific lore, suggestions for settlers, bits of local history, and much general information about the flora and fauna of the new country. In regard to climate, he wrote:

Cold weather begins with the middle of November, the winter's perpetually freezing, insomuch that their Rivers and salt-Bayes are frozen over and passable for Men, Horse, Oxen and Carts: Aequore cum gelido zephyrus fert xenia Cymbo. The North-west wind is the sharpest wind in the Countrie. In England most of the cold winds and weathers come from the Sea, and those seats which are nearest the Sea-coasts in England are accounted unwholsome, but not so in New-England, for in the extremity of winter the Northeast and South-wind coming from the Sea produceth warm weather, only the North-West-wind coming over land from the white mountains (which are alwayes [except in August] covered with snow) is the cause of extream cold weather, always accompanied with deep snowes and bitter frosts, the snow for the most part four and six foot deep, which melting on the superficies with the heat of the Sun, (for the most part shining out clearly every day) and freezing again in the night makes a crust upon the snow sufficient to bear a man walking with snowshoes upon it. And at this season the Indians go forth on hunting of Deer and Moose twenty, thirty, forty miles up into the Countrie.

Thomas Higginson wrote home in 1630: "a sup of New England air is better than a whole draught of Old England ale."

Restored windmill on Nantucket Island. Archives of the Nantucket Historical Association.

New Sweden on the Delaware

The distinction of being America's first weatherman is generally accorded the Reverend John Campanius, or John Campanius Holm (of Stockholm), who lived from 1601 to 1683. Certainly he was the first resident of what became the United States to keep a systematic daily record of the weather over a substantial period of time. Today our National Weather Service honors his memory with the John Campanius Holm Award, given annually to deserving volunteer weather observers for long, dutiful service to the scientific interests of their country.

Campanius arrived in New Sweden on the Delaware River in February 1643. For at least two years, 1644 and 1645, he maintained his daily weather record. Though he did not have any scientific instruments, his daily jottings give an informative account of the climatic pattern along the Delaware River, and his notes, edited by his grandson, were published in Sweden in 1702:

> The climate and temperature in Virginia and New Sweden are variable, as with us; some years are colder, and some years are warmer . . .
>
> The severity of winter lasts, at most, two months; it begins in January, when it is somewhat cold, and then it increases, so that before Christmas there is very little cold, but only wind and rain: the end of January, and the beginning of February are the coldest parts of the winter.
>
> Spring is very fine and pleasant, without any stormy or rigidly cold weather, but only small soft rains and a clear sky.
>
> The summer is, for the most part, pleasant and moderately warm, except in August and September, which are the hottest parts of the year, and in some years it is so warm that people long for rain and wet weather, by which the air is immediately cooled.
>
> The autumn is pleasant and dry, and sometimes a little cold, as was observed by John Campanius in the year 1645.

Another early description of the Delaware climate came from the pen of Peter Lindstrom, an engineer serving in the colony. It shed little light on actual weather conditions, but was climatologically quaint, at least in translation, as the following passage shows: "It doth not often rain, but when it does, it is generally with lightning and thunder, tremendous to the sight and hearing. The whole sky appears to be on fire, and nothing can be seen but smoke and flames."

Virginia

The climate of the tidewater peninsula lying between the York and James rivers, where the Jamestown Colony was founded in 1607, played a dominant role in compounding the difficulties that beset its first year of settlement. A long hot summer with enervating heat was followed by a warm autumn with attendant death-dealing fevers. Many colonists, already weakened by the usual ills experienced on a long sea voyage in those days, fell victim to those twin scourges of heat and fever. Then came a harsh winter, apparently one of the severest of the century. Historian William Strachey related: "In the year 1607–08 was an extraordinary frost in most of Europe and this frost was as extreme in Virginia." Captain John Smith in his *General History* referred to "the extreme frost of 1607" and "the extremity of the bitter cold." George Percy, author of the only semblance of a diary kept during the first year, blamed the severe cold as the chief cause of the deaths of more than half of the colonists during the winter. Of the 105 persons comprising the original colony, only 32 survived by the time the first supply ship arrived from England.

The first descriptive account of the climate of Virginia was published in 1612 by Captain John Smith in *A Map of Virginia*, and practically the same wording appeared in the contemporary *The Historie of Travaile into Virginia Britannia* by Strachey:

> The temperature of this country doth agree well with English constitutions, being once seasoned to the country; which appeared by this, that by many occasions our people fell sick; yet did they recover by very small means, and continued in health, though there were other great causes, not only to have made them sick, but even to end their days.
>
> The summer is as hot as in Spain; the winter cold as in France or England. The heat of summer is in June, July, and August, but commonly the cool breezes assuage the vehemency of the heat. The chief of the winter is half December, January, February and half March. The cold is extreme sharp, but here the proverb is true, that no extreme long continues.

The Development of Climatological Knowledge

At the commencement of the eighteenth century all of the colonies that were to form the beginnings of the United States of America had been planted with the exception of Georgia. Yet during the first hundred years of settlement only the coastal fringe of the continent had been explored and its geography and climate assessed by the English. The French fathers and voyageurs had followed the convenient network of waterways into the heart of the continent by way of the Great Lakes and the Mississippi Valley, yet the communication of their intelligence was to the Old Continent and not to the English settled from Maine to Carolina.

The eighteenth century would witness the penetration of the geographical and political barriers of the Appalachians and the removal of the Indian menace east of the Mississippi River. Toward the end of the century, thermometers and barometers were carried to such formerly remote places as Natchez, St. Louis, and Detroit; and by 1800, Kentucky and Ohio were in the process of receiving substantial waves of immigrants. The trend of the seasons in this vast region became known through the correspondence of settlers and the diaries of explorers and travelers.

During the first decades of the nineteenth century, great arguments arose about the climate of the Great Plains, which then lay beyond the fringe of current settlement. Was there a "Great American Desert" forever barring the type of agricultural settlement known in the eastern woodlands? This question remained unresolved through midcentury. The main source of climatological information came through the westward push of U.S. Army detachments to establish forts to contain the Indian menace and to protect the principal routes of communications following the conclusion of the War of 1812. It was the duty of the Medical Department surgeon at each post to keep a "Diary of the Weather," with a view to compiling a "medical topography" of the region. Though of unequal quality, the composite of the observations did give an indication of the general climate conditions prevailing west of the Mississippi River. Beginning in the late 1840s, the military weather records were supplemented by the reports of volunteer observers of the Smithsonian Institution in Washington, which supplied an array of thermometers, barometers, and rain gauges to qualified individuals.

By 1850, after thirty years of record-collecting, enough data had been accumulated so that an analysis could be made and a coherent presentation assembled of the actualities of the American climate. The task fell to a youthful statistician, Lorin Blodget of Pennsylvania, whose masterful production won him the well-deserved title of "father of American climatology." With the publication in 1857 of his four-hundred-page volume, *Climatology of the United States and the Temperate Latitudes of the North American Continent*, the general features of our climate became part of man's knowledge.

Impact of the
New World Climate

"It would be impossible, indeed, to cross an ocean anywhere else to find so little that is unfamiliar on the opposite side," Professor Carl Sauer, a founder of the American school of historical geography, has written of the western Europeans who crossed the Atlantic in the seventeenth century to find homes in the New World.

The basic geological similarities of northwest Europe and northeast North America have provided strong evidence to support the hypothesis of continental drift, which postulates that the two landmasses were once joined. The flora and fauna also show many similarities. Professor Sauer had these mainly in mind when composing his study of the familiar plant types and animal life the Western Europeans found in the temperate zone of North America.

In the realm of climate, however, his general statement needs some modification. Though the lands of New France were in the same latitude as central France, the temperature and precipitation differed markedly. Furthermore, the latitude of New England stood ten to twelve degrees south of Old England, but the assumed amelioration of climate extremes did not follow. In fact, winters were more severe than anticipated, and summers much hotter than anything experienced in the homeland. The region of the new colonies was subject to continental controls rather than to the maritime influences that predominate in the British Isles.

The following tables illustrate the climatic differences be-

	LONDON	BOSTON	NEW YORK	NORFOLK	CHARLESTON
Temperature Means in Degrees Fahrenheit					
Coldest month	39.8° January	29.2° January	32.2° January	40.5° January	48.6° January
Warmest month	63.7° July	73.3° July	76.6° July	78.3° July	80.2° July
Annual	50.6°	51.3°	54.5°	59.3°	64.7°
Precipitation Means in Inches					
Wettest month	2.53 November	4.51 November	4.01 August	5.92 August	8.21 July
Driest month	1.54 March	2.74 July	2.85 October	2.71 April	2.13 November
Annual	24.00	42.52	40.19	44.58	52.12

tween London and the principal areas of early settlement. Temperature means are given for the coldest and warmest months of the year and the annual mean. Precipitation amounts are for the wettest and driest months of the year and the annual mean precipitation.

The period of settlement of the original American colonies must be considered in relation to the climatic changes taking place in the northern hemisphere over the past millennia. Since the final glacial retreat from the mainland of North America about 10,000 years ago, several swings of the climatic pendulum have brought ever-changing temperature conditions. From a cold period at the end of the Middle Ages in the 1300s that firmly closed the sea access to Greenland and sometimes to Iceland, a brief warming period followed in the late 1400s and 1500s. Then a distinct thermal reversal followed, and the coldest temperatures since glacial times prevailed from the early 1600s to almost the end of the 1800s. Paleoclimatologists aptly named this the "Little Ice Age" to dramatize it as a period when glaciers advanced in Europe and probably in North America to positions approaching their former moraines, and the boreal forests on both continents retreated southward into the temperate zone.

No worthwhile instrumental measurements of temperature were made in critical places during the seventeenth century, but indirect evidence has been drawn from pollen deposits in ponds and bogs, ocean bed cores, ice layers, tree rings, and glacial behavior. All point to the basic fact that, at the time of the founding of the permanent colonies along the Atlantic seaboard, atmospheric conditions were somewhat harsher than they are now. Temperatures ran one to two degrees Fahrenheit cooler, and the growing season was shorter by one to two weeks. Snows and frosts came earlier in the autumn and continued later in the spring.

Outstanding Colonial Weather Events

Seventeenth Century

1604–05   First wintering-over of permanent settlers from Europe within the present boundaries of the United States; severe conditions caused French to move in spring from the St. Croix River settlement on an island between the present State of Maine and the Province of New Brunswick to a milder maritime exposure at Annapolis, Nova Scotia.

1607–08   Settlers of the Virginia Colony at Jamestown endured "extreme frost"; only 32 of 105 survived by spring; colony saved by timely arrival of supply ship from England when about to be abandoned.

1620–21   Pilgrims' first season at Plymouth was described as "a calm winter such as never seen here since"; only substantial snowstorm came in December; some snow and sleet fell in mid-February; otherwise rainstorms and mild temperatures prevailed.

1629–30   Puritans' first winter on Massachusetts Bay described as "very mild season, little frost, and less snow, but clear serene weather, few northwest winds," by Thomas Higginson, *New Englands Plantation*.

1635   New England, Aug. 15: The Colonial Hurricane, a tropical storm of extreme force, equal to Sept. 1938, cut across southeast Massachusetts and Rhode Island; caused tide to rise 14 feet, drowning Indians; unroofed houses and blew down thousands of trees.

1637–38   "This was a very hard winter," wrote Gov. John Winthrop in his *Journal*. "The snow lay, from November 14th to April 2nd, half a yard deep and a yard deep beyond Merrimack [River], and so the more north the deeper, and the spring was very backward." Boston Harbor froze over solid; snow fell on May 3.

1638 Boston area, Aug. 13: "In the night was a very great tempest, or huricano at S.W. which drave a ship on the ground at Charlestown, and brake down the windmill there, and did other harm," Winthrop *Journal*.

1638 New England, Oct. 5: ". . . was a mighty tempest, and withal the highest tide, which has been since our coming into this country," Winthrop *Journal*; in Maine, "The greatest mischief it did us was the wracking of our shallops, and the blowing down of many trees, in some places a mile together," John Josselyn, *Account of Two Voyages to New-England*.

1641–42 New England: "The frost was so great and continual this winter, that all the bay was frozen over, so much and so long, as the like, by the Indians' relation, had not been these 40 years, and it continued from the 28th of this month [January] to 3rd March," Winthrop *Journal*.

1643 Massachusetts, July 15: "There arose a sudden gust at N.W. so violent for half an hour, as it blew down multitudes of trees. It lifted up their meeting house at Newbury, the people being in it," Winthrop *Journal*.

1644–45 New England: "The winter was very mild, and no snow lay, so as ploughs go most of the winter," Winthrop *Journal*.

1646–47 Roxbury, Mass.: "This winter was one of the mildest we ever had; no snow all winter, nor sharp weather. We never had a bad day to go preach to the Indians [at Natick] all this winter. Praised be the Lord," the Rev. John Eliot.

1665–66 Boston: "The first week of January the frosts were violent, Charles River was passed on foot, and only the channel open before Boston," John Hull *Notes*.

1667 Virginia, Sept. 6: "Strange News from Virginia, being a true relation of the great tempest in Virginia"; this pamphlet described a strong hurricane in Virginia that overturned houses, stripped fields of crops, and raised a tide of 12 feet above normal; high floods on rivers followed.

1675 New England, Sept. 7–8: Mature hurricane of great force swept across southeast, considered equal to 1635; blew down windmill at Newport; damaged ships at Boston; caused "much loss of hay and corn. Multitude of trees blown down," John Pike *Journal* at Dover, N.H.

1677 Plymouth, Mass., Dec. 14: "Such a Dreadful storme, as hath not bin knowe these 28 years," the Rev. John Cotton *Diary*.

1680 Cambridge, Mass., July 18: Tornado described by eyewitness, the Rev. Increase Mather in *Remarkable Providences*; it tore up trees by the roots and carried away large stones; one person killed.

1680–81 New England: Severe winter; "Hath been a very severe winter for snow and a constant continuance of cold weather; such as most affirm hath not been for many years," Increase Mather; William Penn at Philadelphia commented that it was the severest known to anyone living there.

1682 Connecticut, June 18: Major tornado swept through Stratford, Milford, Fairfield, and New Haven and disappeared in Long Island Sound; three barns and one house blown down.

1682 Springfield, Mass., July 6: Large chunks of ice fell from sky, some nine inches in diameter; shingles broken, roofs holed.

1683 Boston area, Feb. 19: Warm rain fell on deep snow cover, causing high floods on streams; many bridges and dams carried away.

1683 Atlantic coast, Aug. 23: Hurricane struck from Virginia to Massachusetts; extensive damage in Rhode Island; torrential rains raised Connecticut River 26 feet above usual level; crops inundated and carried away.

1692 Hartford, Conn., Feb.–Mar.: Great flood on Connecticut River; stage said to have risen to 26 feet, highest until 1801.

1693 Virginia to Long Island, Oct. 29: Acomac Hurricane; altered shoreline, especially on Acomac [Delmarva] peninsula; Fire Island inlet on Long Island opened; some inlets closed; ships driven ashore.

1695 Boston, Apr. 29: "A very extraordinary storm of hail, so that the ground was made white with it . . . 'twas as big as pistol and musquet bullets," Judge Samuel Sewall *Diary*.

1696 Salem, Mass., Sept. 25: "A black frost. Ye ice on ye side of my house as thick as window glass," John Higginson.

1697–98 New England and Middle Colonies: "The terriblest winter for continuance of frost and snow, and extremity of cold, that ever was known," Sudbury, Mass., record. Judged the severest of the century and not equaled until 1740–41.

1699 Boston, Dec. 10: "The rain freezes upon the branches of the trees . . . considerable hurt is done in orchards," Judge Samuel Sewall *Diary*.

1700 South Carolina, Sept. 14: Hurricane struck ship *Rising Sun* lying at anchor offshore; 97 Scottish immigrants drowned; houses in Charlestown overturned and town flooded.

Eighteenth Century

1704–05 Severe winter; series of storms left snow almost three feet deep at Philadelphia in late January; great storm on Jan. 26 caused highest tide in 20 years at Salem, Mass.; hard freeze at Boston as late as Apr. 23.

1705–06 Second severe winter in row; cold at Christmas 1705 froze Hudson River at New York City; 132 sailors froze to death when ship grounded on Sandy Hook in bitter gale; winter began Nov. 30 and lasted until Feb. 13 at Dover, N.H.; cold, backward spring followed.

1713 Charleston, S.C., Sept. 16–17: Fortifications damaged by hurricane winds, but they saved town from tidal inundation; ship driven three miles into woods; houses undermined and destroyed in North Carolina; 70 persons drowned.

1716 Massachusetts, Oct. 24–25: hurricane swept offshore shipping, dismasting several ships; noticed at Cape Cod,

Martha's Vineyard, and Boston area; trees and fences blown down.

1716   New England, Nov. 1: Dark day caused by smoke in atmosphere; half hour darkness during Sabbath services thought to be an omen of God's displeasure.

1717   New England, Feb. 27–Mar. 7: The Great Snow of 1717, the most celebrated snowstorm in colonial history; consisted of two major and two minor snows over a nine-day period; accumulated to 36 inches at Boston, 60 inches to the northward; all travel, even church-going, halted for two weeks.

1717   Cape Cod, Apr. 27: Pirate ship *Whidnah* wrecked in storm with loss of 144 crew members.

1719–20   Atlantic seaboard: Severe winter, coldest since 1697–98 at Boston; Hudson River at New York City crossed on foot in mid-January; sheep buried alive in snow in Boston area; said to have been coldest ever known in Virginia.

A

Difcourfe

OCCASIONED

By the late D.ftreffing

STORM

Which began *Feb.* 5*eth.* 1716, 17.

As it was Deliver'd *March*

3*d.* 171⁶⁄₇.

By **Eliphalet Adams**, A.M.

Paftor of the Church in New London.

Pfal. cxlviii. 8.

Fire and Hail,, Snow and Vapours, Stormy
Wind fulfilling His Word.

NEW-LONDON:

Printed and Sold by *T. Green,* near the
Meeting-Houfe. 1717.

Opposite: Great snow in 1717. John W. Barber, The History and Antiquities of New England *(Worcester, Massachusetts: Dorr, Howland, and Co., 1841).*

. . . and sermons were published about the great snowstorm of 1717.

1722   New Orleans and Mobile, Sept. 23–24: First recorded hurricane on middle Gulf Coast; 34 houses destroyed at New Orleans, including church, parsonage, and hospital; every ship in port damaged; tide rose 8 feet in Bay St. Louis in present Mississippi.

1723   Boston, Feb. 23: Severe northeaster during full moon raised greatest tide ever known by 20 inches; streets and cellars near harbor flooded; described by the Rev. Cotton Mather in letter to Royal Society and in humorous article by Benjamin Franklin in *New England Courant.*

1724   Pennsylvania, Aug. 14: tornado pursued course from Chester County into Bucks County, just northwest of Philadelphia; windfall through woods; barns unroofed; mill destroyed.

1724   Virginia, Aug. 23: "The Great Gust of 1724"; "We have had such a violent flood of rain and prodigious gust of wind that the like I do not believe ever happened since the universal deluge," John Custis letter; most of tobacco destroyed along with some houses and vessels.

1727   New England, Sept. 27: Hurricane of 1727; "then the Lord sent a great rain and horrible wind; whereby much hurt was done, both on the water and on the land," the Rev. Samuel Phillips in sermon; many ships at Marblehead wrecked; trees torn up by roots; chimneys downed at Boston; barns demolished.

1728   Carolinas, Aug. 13: hurricane caused damage in Charleston harbor; 23 ships damaged or lost; houses, wharves, bridges destroyed; great crop losses; trees along coast leveled.

1732–33 Atlantic seaboard: severe winter with most harbors frozen over; "one of the most severe in Chesapeake history," Arthur P. Middleton history; Casco Bay in Maine solid for two months; Delaware Bay did not open until Mar. 15; Feb. 25 and 26 were "coldest ever felt at Boston."

1740   New England, mid-December: breakup of early winter caused floods on all rivers; "highest ever known" on Merrimack in New Hampshire; greatest since 1692 on lower Connecticut River.

1740–41 Atlantic seaboard: "The Hard Winter of 1741"; severest since 1697–98; cold, snowy November, followed by thaw and floods in early December; great snowstorm on Dec. 28; thaw in late January; very severe conditions in February; Hudson River at New York City shut on two occasions; snow three feet deep in central Connecticut; rivers continued frozen until mid-April.

1743   Atlantic coast, Nov. 2: Ben Franklin's Eclipse Hurricane; observations at Philadelphia and Boston enabled Franklin to conclude that northeast storm came from the southwest; hurricane surge caused highest tide at Boston in 20 years; damage to wharves and shipping.

1745   Charleston, S.C., Jan. 10–11: "Sudden change" in temperature — dropped from 70°F (21°C) to 26°F (−3°C), and then sank to 15°F (−9°C) next morning; "the greatest and most sudden change I have seen," Gov. James Glen.

1747–48 New England: "The Winter of the Deep Snow"; "we had about 30 snows and less thawing weather than usual, so that the snow lay upon the ground till it came to be 4 or 5 feet deep," First Church of Cambridge, Mass., records;

snow cover 2 to 3 feet deep as late as Apr. 15.

1748   Southern Colonies, Feb. 17: Coldest Colonial Day; "at 8 o'clock, it was at the tenth degree and no doubt had been lower before that, as the spirits in the thermometer were then rising, the air being warmed by the sun," Gov. James Glen.

1749   Eastern New England, May–June: Severe drought; fast held throughout province on June 15 to pray for rain; "fields as dry and almost white as in winter time"; rains came after July 1, reviving crops; Day of Thanksgiving held on Aug. 27 at Salem.

1749   Middle Colonies and New England, June 29: Famous Hot Sunday; Benjamin Franklin's thermometer reached 100°F (38°C); "a hot, burning season" in Massachusetts; young Ezra Stiles preached his first sermon that day.

1749   Middle Atlantic Coast, Oct. 18–19: Strong hurricane caused havoc among coastal shipping; 8 vessels lost on Ocracoke Island; tide in Chesapeake Bay rose 15 feet, inundating Norfolk and Annapolis; 7 ships ashore on Martha's Vineyard.

1752   South Carolina, Sept. 15: "The Great Hurricane of 1752"; most destructive in Charleston's history; tide rose 10 feet above normal high water, bay and river almost joined; shift of wind with passage of center saved city from complete inundation; all ships but one in harbor driven ashore; plantations suffered severely.

1752   South and North Carolina, Oct. 1: Second severe hurricane struck main blows between Cape Fear and Cape Lookout; Onslow County Court House destroyed with all records; heavy rains in Virginia and Maryland.

1752–53⎫ ⎰Atlantic Coast: Series of four mild winters with light
1755–56⎭ ⎱snow and little cold; considered remarkable at the time.

1756   Ohio River, spring: Great flood nearly carried away newly established Fort Duquesne at present site of Pittsburgh.

1756   Middle Colonies, June 22: Squall line caused strong wind rush over Maryland, Delaware, Pennsylvania, New Jersey, and New York; several tornadoes did structural damage; 200 houses blown down in St. Marys County, Md.; tornadoes hit Jamaica on Long Island and in Essex County, N.J.

1757 New England, Sept. 23–24: Offshore hurricane did much damage to shipping; later hit Grand Banks fishing fleets.

1759 Leicester, Mass., July 10: Tornado with 6-mile path; struck house full of people, injuring most; Negro killed after being carried through air; described by Prof. Winthrop of Harvard.

1760 Boston, Mar. 20: Great Boston fire; strong northwest wind carried flames across northeast part of city, sparing rest.

1761 Charleston, S.C., May 4: Tornado passed down Ashley River creating waterspout; hit fleet assembling for voyage to England; three ships sunk, others damaged.

1761 Southeast New England, Oct. 23–24: Strong hurricane brushed Rhode Island; windmill and church steeple blown down at Newport; worst blow at Boston since 1727, when trees torn up by the roots.

1762 Ohio River, Jan. 9: High flood reached 39 feet above low water; houses and stores carried away.

1762 New England, late-spring and summer: Severe drought; very little rain fell from Apr. 20 to Aug. 30; "everything appeared to have been burned" in the fields; public fasts held; very poor harvests and hay scarce, sold for four times usual prices; cattle slaughtered for lack of fodder; fires in fields burned some dwellings.

1763 Ohio River, Mar. 9: Second high flood in two years raised river stage higher than in 1762; extensive damage.

1764–65 Northeast: Severe winter; New Year's cold spell lowered thermometer to 0°F (−18°C) at New York City freezing Hudson River; second cold spell at end of January sent mercury to −6°F (−21°C) at New York, to −9°F (−23°C) at Cambridge, Mass.; Delaware River frozen; two feet of snow at suburban Germantown.

1769 Carolinas to New England, Sept. 8–9: Powerful hurricane crossed eastern Carolinas; blew down thousands of trees, demolished Brunswick County Court House; Newburn inundated; crops ruined; heavy damage to shipping northward along coast; tree destruction in Philadelphia, New York City, and Boston.

1770 New England, Jan. 6–7: Heavy rainstorm fell on deep snow cover and caused widespread floods; many bridges and mill dams carried away.

1770 New England, Oct. 20: Late Season Hurricane of 1770; struck Rhode Island and eastern Massachusetts; highest tide since 1723 inundated Boston streets; some structural damage on land.

1772 Virginia and Maryland, Jan. 26–29: Washington & Jefferson Snowstorm; "three feet deep everywhere" around Mount Vernon; 22 inches reported by Jefferson at Monticello; 33 inches measured at Winchester, Va.

1772 New England, Mar.–Apr.: Series of heavy snowstorms caused record deep snow cover; final big storm on Apr. 2–3 left 2 feet on ground at Boston, 3 feet near Hartford, Conn.

1773 New England, Feb. 21: "Memorable Cold Sabbath"; northwest gales drove thermometer below zero; people froze extremities going to church; −10°F (−23°C) at Salem, Mass., −20°F (−29°C) at Portsmouth, N.H.

1773 Carolinas, Georgia, Florida, Feb. 21–22: 19°F (−7°C) at Charleston with heavy snowfall, "greatest ever known"; ship off St. Augustine, Fla., reported heavy fall of snow, later referred to euphoniously as a "white rain."

1775 Atlantic seaboard, Sept. 2–3: Independence Hurricane; strong tropical storm took inland path from North Carolina to central Pennsylvania; southeast winds caused highest tide known on Delaware River; much damage to shipping.

1775 South Carolina and Georgia, Dec. 23: The Snow Campaign; snowfall of 18 to 24 inches impeded soldiers in first skirmishes of war in South; cold, snowy winter followed.

1776 Virginia, Dec. 25–26: Heavy snowstorm dropped 22

inches at Monticello; over a foot at Moravian settlements in North Carolina; fringe of storm caused snow and sleet at Battle of Trenton.

1778 Gulf Coast, Oct. 7–10: Strong hurricane struck from Louisiana to Florida; bayou settlements inundated by high water; all ships but one sank at Pensacola.

1778 Cape Cod, Nov. 3: H.M.S. *Somerset* grounded in storm; crew of 480 captured and marched to Boston as prisoners of war.

1778 New England, Dec. 25–26: Hessian Storm raged with blizzard conditions; several German mercenaries froze to death near Newport; zero temperatures and 18 inches of snow at Boston.

1779 Louisiana, Aug. 17–18: Hurricane at New Orleans damaged houses and caused shipwrecks; William Dunbar detected central eye of storm with circulatory winds.

1780 New Jersey to Maine, May 19: The Dark Day; smoke from forest fires aloft filtered sun's rays, producing dark atmosphere all day and giving foliage a brassy hue.

1780 Gulf of Mexico, Oct. 18–21: Solano's Hurricane; fleet under Spanish admiral prevented from attacking British

May 19 brought the "wonderful dark day" in 1780. Richard M. Devens, Our First Century *(Springfield, Massachusetts: Nichols, 1880).*

Pensacola by severe tropical storm that scattered the ships.

1780–81 Eastern United States: "The Hard Winter of 1780"; rated the severest in American history; series of heavy snowstorms in December and early January accumulated 3 feet of snow in New York area and 4 feet in New England; severely cold January; all harbors frozen from North Carolina to Maine; most travel by land and water halted; great economic distress prevailed.

1782 Louisiana, spring: Highest flood known in the Mississippi River inundated many settlements.

1783 Atlantic Coast, Oct. 7–9: Hurricane swept from Carolinas to New England; shipping battered; heavy rains fell inland; raised highest tide known in 40 years at New Haven.

1783 Northern New England, Oct. 22–23: Heavy rain fell on deep snow cover in Green and White mountains producing "the greatest flood ever known since the settlement of this country," according to the *Vermont Journal*.

1784 Louisiana, Feb. 13: Mississippi River at New Orleans filled with ice; on 19th the ice disappeared before city, passed downstream and out into Gulf of Mexico as ice floes; only other similar instance occurred in February 1899.

1784 Hartford, Conn., Feb. 10–17: Very cold week, minimums ranging from −12°F to −20°F (−24°C to −29°C) on eight consecutive mornings; New York Harbor froze at the Narrows; deep snow cover general.

1785 Carolinas and Virginia, Sept. 23–24: Severe hurricane on Outer Banks of North Carolina, extensive erosion; shipping damaged in Chesapeake Bay.

1785 Northeast, Oct. 23–24: Great rainstorm caused high floods; 9 inches of rain raised Merrimack River in New Hampshire to highest level known; Coos Meadows at Newbury, Vt., inundated.

1786 Northeast, Dec. 5–6: Heaviest snowstorm since 1717 in New England followed by another major storm on Dec.

9–10; New Haven reported 37 inches total; highest tide ever known did extensive damage at Nantucket.

1787   Central New England, Aug. 15: Four-State Tornado Swarm in Connecticut, Rhode Island, Massachusetts, and New Hampshire; destroyed much property, though no large settlement hit.

1788   Virginia, July 23–24: George Washington's Hurricane; described by George Washington in diary as center passed over Mount Vernon; Norfolk, Annapolis, and Chesapeake Bay communities badly battered.

1788   Northeast, Aug. 19: Hurricane with small central core did enormous forest damage from New Jersey northeast over Hudson Valley and through north-central New England; several persons killed by falling trees.

1788   New England and Middle States, Feb. 5: Cold day, New Haven maximum only + 3°F (− 16°C); at Philadelphia the mercury fell from 38°F (3°C) to a minimum of − 6°F (− 21°C); called "Cold Tuesday" at Morristown, N.J.

CHAPTER TWO

The Field
of Battle

It is not very flattering to that glory-loving, battle-seeking creature, Man, that his best-arranged schemes for the destruction of his fellows should often be made to fail by the condition of the weather. More or less have the greatest of generals been "servile to all the skyey influences." Upon the state of the atmosphere frequently depends the ability of men to fight, and military hopes rise and fall with the rising and falling of the metal in the thermometer's tube. Mercury governs Mars. A hero is stripped of his plumes by a tempest, and his laurels fly away on the invisible wings of the wind, and are seen no more forever. Empires fall because of a heavy snow. Storms of rain have more than once caused monarchs to cease to reign.
— Charles C. Hazewell, *Atlantic Monthly*, May 1862

War of American Independence

The War of American Independence marked another bloody episode in the century-long struggle between the British and French empires for world domination, and the events of 1775 and 1776 introduced new elements into the conflict. A period of intense turmoil and upheaval commenced among the peoples and governments of Europe as a result of the new philosophical ideas formulated by the political thinkers of the age and put into a practical document by the American revolutionists. It was the century of the "Enlightenment," when reason was first turned to the task of explaining in nonreligious terms the nature of man and in scientific terms the behavior of the natural world.

At the time of the outbreak of hostilities between the British Crown and the American colonists, the pursuit of serious meteorological studies was about a century old. The main progress had been in the development of satisfactory instruments with which to make the basic atmospheric measurements. A workable mercury thermometer had been designed and manufactured, though there was no uniformity in construction or standard of agreement in readings. Thermometers were generally in the hands of "gentlemen" of the educated class: professors, ministers, lawyers, and plantation owners. From mention of readings in colonial newspapers and in diaries of individuals, we can estimate the presence of about twenty-five thermometers in the colonies, although the number may have been double that figure.

There were barometers, too. Consisting of a long glass tube filled with mercury, they were fragile and hard to transport; and even if safely shipped, they were difficult to adjust for accurate readings and almost impossible to calibrate for altitude. They were usually part of a dual instrument housed in a wooden frame with a thermometer attached. Their exposure was far from ideal.

Weather instruments were in the hands of individuals. Although suggestions were made for the coordination of observations and the exchange of data between men of scientific bent, nothing was accomplished along these lines. There was no government activity in this field.

Little was understood about the movement of atmospheric systems or their internal structure and behavior. What forecasting there was came from the application of time-honored proverbs and folklore handed down by classical authors and repeated in the almanacs of the day. An individual was limited to his own horizon. He could read his instrument and study the signs in the sky; these were his sole aids in making a forecast for local use only, since no means of rapid communication of weather reports existed.

Sources of information about weather conditions attending the battles of the War of Independence are rather limited. Fortunately, the records of Professor John Winthrop at Cambridge have been preserved in the Harvard Archives. These cover the years through 1779 and supply daily details for the Boston area in 1775–76. Also valuable are the records of Dr. Edward Holyoke at Salem, which cover the entire war period. President Ezra Stiles of Yale College maintained daily records during the war period, mainly at New Haven. No record for the New York City area for these years has been located. The British Army did have a thermometer at its headquarters on Manhattan Island throughout the occupation, but no continuous record of its readings is known to exist. A member of the American Philosophical Society kept a daily record at Philadelphia from 1775 through May 1778, which covers the Philadelphia Campaign and the winter at Valley Forge.

The only true meteorological observer in the South was the Reverend James Madison at William and Mary College. Only fragments of his observations remain for Williamsburg because his personal papers were destroyed when his home burned in 1781. Thomas Jefferson was on the move during the war years and did not keep his meteorological records as he would in later years. For the Carolinas and Georgia we have nothing of a scientific nature, though we know of thermometers at New Bern, Charleston, and Savannah prior to the outbreak of war.

To supplement the few meteorological records, numerous diaries of combatants mention the weather, some almost daily, others only occasionally, producing valuable background material for judging the weather attending some of the major battles. Fortunately, most of these records, however brief, have been published by proud ancestors or patriotic historical societies.

Thus, from a composite of meteorological records, personal diaries, and newspaper accounts, one can assemble a reasonable description of the weather conditions attending most of the principal battles of the war, except for those fought in the Carolinas and Georgia.

Lexington and Concord:
An Ideal Day for Outdoor Activity

Weather conditions on the day of the military engagements at Lexington and Concord were long a subject of curiosity among historians. This topic was discussed at several meetings of the Massachusetts Historical Society in the nineteenth century, but without the introduction of much positive evidence. Several diarists were quoted as attesting that the weather was "cool" — contrary to a long-standing tradition that the day had been "warm for the season" and that "early fruit trees were in blossom."

At least three monographs on the military aspects of the Lexington and Concord action were produced by scholars in the Boston area in anticipation of the 150th anniversary celebration, but these shed little factual light on the weather for April 19, 1775. Apparently none of the authors extended his research into the readily available Harvard University Archives where the meteorological records of Professor John Winthrop, a sixth-generation descendant of the founder and first governor of the Massachusetts Bay Colony, have been carefully preserved in his original notebooks — which contain thermometer, barometer, and rain-gauge readings for almost every day from 1742 until his death in 1779.

Winthrop's records indicate that a cold front passed through eastern Massachusetts about noon on April 18, bringing an end to the showery conditions, a shift of wind to the west, a rising barometer, and rather rapid post-frontal clearing. Visibility was good late in the evening, when the signal lamps were hung in the steeple of Old North Church and seen by Colonel Conant at Charlestown across the river. Paul Revere later recalled that "the moon was rising" and that it was "very pleasant" at his zero hour of eleven o'clock. The *Almanack for 1775* by Nathaniel Ames scheduled the moon, now four days past full, to rise at 10:48 P.M.

The weather station location at Harvard College was close to the line of advance and withdrawal of the British regulars. Winthrop's thermometer at 6:00 A.M. on the nineteenth read 46°F (8°C), and his barometer 29.56 inches (100.1 kPa) and rising, wind light from the west, and a "very fair" sky overhead — in all, a cool, fresh spring morning. His observation was made soon after the skirmish had taken place at Lexington Green during the hour after daylight. Sunrise came at 5:19 A.M., according to the Ames *Almanack*.

Winthrop, as was his custom, took his second observation

at 1:00 P.M. after the embattled farmers had made their morning stand at Concord Bridge and while the King's troops were making their painful withdrawal through Lexington and Arlington on the road back to Charlestown. At this time, the wind had picked up to moderate from the west, still bringing cool, dry air across the area. The thermometer rose only to 52°F (11°C), and the sky was "fair with clouds" — probably fair-weather cumulus, floating peacefully across the troubled scene below. Winthrop concluded his observations with the laconic notation: "Battle of Concord will put a stop to observing."

The general conditions existing in the area on Lexington-Concord Day were confirmed by the contemporary meteorological observations of Dr. Edward A. Holyoke at Salem, some 25 miles east-northeast of Concord. The passage of the storm system on the eighteenth was noted, with "moist air and rain" in the morning, followed by "fair with some wind" in the afternoon and evening. Holyoke's morning thermometer reading on the nineteenth was 51°F (11°C), but this was in a sheltered door entry. His remarks for the day were "serene, dry air, pleasant, cool."

For a local view of the weather on the famous day, the diary of the Reverend Jonas Clarke of Lexington noted: "April 18. a fine rain. April 19. clear, windy." Other weather-watchers in eastern New England made diary entries indicating that fine conditions prevailed generally on the nineteenth.

All meteorological evidence indicates that the Day of Lexington and Concord had a crisp morning and a comfortable afternoon — skies were fair and wind movement fresh — in all, an ideal day for outdoor activity.

Bunker Hill Day: Dry Ground and Clear Skies

The American initiative in fortifying the crown of Breed's Hill (adjacent to Bunker Hill) across the Charles River from Boston precipitated one of the bloodiest engagements of the entire War of Independence, considering the relative number of combatants involved. The construction of the colonists' small redoubt was carried out in about four hours, between midnight and dawn on the morning of June 17, 1775.

Two British warships anchored in the estuary between Boston and Charlestown were keeping watch on the enemy, as well as on current weather conditions. H.M.S. *Lively*, whose lookout first sighted the rebel entrenchments, recorded "moderate and fair weather" in her log at 4:00 A.M. H.M.S. *Glasgow* nearby had "fresh breezes & clear" that morning. Unfortunately, the heated action that followed caused the British log keepers to omit any further weather entries that day.

Dr. Edward A. Holyoke at Salem, some 13 miles northeast of the battlefield, was the nearest meteorologist to the scene. He recorded a morning temperature of 64°F (18°C) presumably taken at 8:00 A.M., as was his custom. Professor Winthrop, located at Andover, some 20 miles north and inland, read 61°F (16°C) on his thermometer that morning. Wind was light from the west, shifting to southwest during the day. Holyoke marked the day was "serene, dry air, hot." Both observers recorded afternoon temperatures of 80°F (27°C), though the exact times of the readings were not specified. Probably the maximum at the battle scene was a bit higher, possibly as high as 85°F (29°C).

The last rains in the Boston area had fallen on the fourteenth, according to our meteorological observers. Though no barometer readings were taken during this period by the two nearby observers, a weather-map reconstruction for the morning of the seventeenth would probably have shown the center of a high-pressure area over eastern New England and southward along the Atlantic seaboard. A cold front with showers had moved through on the fourteenth and was now well out into the Atlantic. A northwesterly flow of polar air from Canada followed — the afternoon reading at Salem on the fifteenth, which would approximate the maximum for the day, was only 67°F (19°C). Winds were west on the sixteenth with the afternoon reading at 72°F (22°C) — the weather "very fair" on both the sixteenth and seventeenth as stable air conditions dominated the area.

Sometime during the afternoon of the seventeenth, when the battle was raging, the winds backed farther into the southwest and by next morning into the south, indicating the eastward passage of the high-pressure ridge and a gradual falling off of barometric pressure. The southerly flow overnight brought a tropical air stream into New England with increasing heat as shown by the afternoon readings at Salem of 91°F (33°C) on the eighteenth and 93°F (34°C) on the nineteenth.

As in the case of many other summertime battles fought long ago, the tradition gradually grew that it took place on "a very hot day." A historian in 1788 declared the day of Bunker Hill was "exceeding hot." Another, writing in 1833, told his readers, "The Battle of Bunker's Hill, it will be recollected, was fought on one of the hottest days ever known in this country."

To set the record straight, the engagement now known as the Battle of Bunker Hill was fought on the afternoon of June 17, 1775, under rather optimum conditions for a mid-June day — the ground was dry, the sky was clear, the temperature was warm but not excessively so, and the air was still relatively dry, though growing more humid hourly. If the action had taken place a day or two later, conditions would have been much warmer, more humid, and the atmosphere more uncomfortable for the participants.

Evacuation of Boston: A Southeaster Foils an Amphibious Assault

March 17 is doubly celebrated by those Bostonians whose ancestors were once residents of the British Isles. Yankees rejoice in Evacuation Day and the departure of King George III's troops; the Irish honor the memory of their patron, Saint Patrick. Curiously enough, both have a weather connection. Unfavorable climatic conditions intensified the ravages of the potato blight and drove more than a million Irish to New England in the 1840s. And a severe storm influenced General Howe's decision to give up Boston and seek military fortune elsewhere.

Military activity around Boston was at a stalemate through the winter of 1775–76, until General Washington moved to fortify Dorchester Heights on the night of March 4–5 with artillery that had been sledded to the seacoast from Ticonderoga. From this position just south of Boston, the American guns could menace the British troops in the town and the transports in the harbor. General Howe immediately ordered an amphibious assault for the next morning. But during the evening hours of March 5 there arose "a hurrycane or terrible storm," as Timothy Newell described it. This was a southeaster of gale proportions, and Boston Harbor is most susceptible to winds from that quarter. At Salem, Dr. Edward Holyoke entered in his meteorological diary for the sixth, "S.E. Very high Wind. Stormy. Rain." The Reverend William Gordon thought it "such a storm as scarce any one remembered to have heard."

The wind and waves disrupted Howe's plans, and he called off the attack, blaming the fiasco on "the badness of the weather." The Americans continued to reinforce their position during these hours of grace, and the British, fearful of a repetition of bloody Bunker Hill, made no further move. Instead, Howe decided to evacuate Boston, an operation he had been contemplating for some time because of his supply problems, the hostility of the local population, and the lack of worthwhile military objectives in the hinterland. If the storm of March 5–6 had not occurred, Dorchester Heights might now have a name in New England history alongside Lexington, Concord, and Bunker Hill.

Battle of Long Island:
That Heavenly Messenger, the Fog

General Washington's uncertainty as to the next British move in their attempt to subdue the rebellious colonies was answered early in the summer of 1776, when General Sir William Howe assembled the mightiest expeditionary force ever sent out from the shores of Great Britain up to that time. The rendezvous of some 500 ships carrying 32,000 trained troops and their equipment was New York Harbor, where they established a base on Staten Island, a strategic location for striking either into New Jersey, onto Manhattan Island, or across the Narrows into Long Island. Washington had already initiated steps to fortify strategic points around the Upper Bay, and now sent orders for all available troops to converge on Manhattan Island.

Battle

The first move of the British forces from their base on Staten Island came on the morning of August 22 after the rains attending a severe thunderstorm had passed off to the east. The weather during the day was logged by H.M.S. *Eagle*: "the first part light breezes and fair, middle cloudy, latter fresh breezes & fair."

The landings of about 15,000 men were made on the western tip of Long Island in Gravesend Bay, a short distance southeast of the present Verrazano Narrows Bridge. The American defensive position ran along a low range of hills stretching from Gowanus Bay across the island in an east-northeast direction. Not until the twenty-seventh did the British attack, and then employed a flanking movement on the American left, which had no natural anchor.

After some rain in the night, the sun rose on the morning of the twenty-seventh in a clear sky — it was "a fine morning" to Captain Bamford. During the late morning military engagement conditions continued favorable. "It was a clear, cool and pleasant day," in the words of Ambrose Serle with the British headquarters. H.M.S. *Roebuck*'s log gave the weather: "first part moderate and cloudy, latter a little wind and fair."

The most influential weather factor that day was the presence of a steady north wind; together with an ebbing tide, this prevented the British warships from moving from New York Bay into the East River, where they could pour their fire into the rear of the American position and cut off the only available escape route to Manhattan Island on the west bank. If the wind had permitted the British to seize control of the East River on this crucial day, "the Revolution might have ended then and there," Christopher Ward, a historian of the campaign, has concluded in *The War of the Revolution*.

After the quick American rout on the twenty-seventh, the British consolidated their forward positions next day, "the morning being very pleasant," but they made no attempt to storm the Americans' last defenses. Then the weather elements made their presence felt again: "In the afternoon we had extraordinary heavy rains and thunder . . . it rained prodigiously," observed the Reverend Shewkirk, "it was awful. The very heavy rain, with intermixed thunder, continuing for some hours till toward evening."

The twenty-ninth brought more "thick weather." H.M.S. *Rose* experienced "little wind and cloudy with hard showers of rain" in the morning, and "light airs and foggy" at the noon hours. In the afternoon "such a heavy rain fell as can hardly be remembered." Ambrose Serle, within the British lines, noted that "a fog and mist obscured the fleet." All reports agreed on the intensity of the rainfall that day.

General Scott wrote of the effect of the storm in the entrenchments before Brooklyn on the twenty-ninth: "You may judge our situation, subject to almost incessant rains, without baggage or tents, and almost without victuals or drink, and in some parts of the lines the men were standing up to their middles in water." Captain Olney observed that "the rain fell in such torrents that the water was soon ankle deep in the fort. Yet with all these inconveniences, and a powerful enemy just without musket shot, our men could not keep awake." Private Philip Fithian added his comments: "The weather is most unfavorable, very rainy; yesterday and to day, so much that Trenches, Forts, Tents, & Camp are overflowed with water, & yet our Men must stand exposed themselves & flintlocks to it all."

Realizing the precarious position of his army in face of a strong foe and the adverse elements, Washington gave orders to assemble boats of all types for the evacuation of his men across the treacherous East River during the night of August 29–30.

Evacuation

The Reverend William Gordon, in gathering material in the 1780s for his history of the war, always exhibited a keen sense of the importance of the weather factor in military affairs. When visiting New York after the peace, he made a study by personal inquiry of conditions attending the evacuation of Long Island and composed an informative account:

> Meanwhile, about nine, the tide of ebb made, and the wind blew strong at northeast, which adding to the rapidity of the tide, rendered it impossible to effect the retreat in the course of the night . . .

But about eleven the wind died away, and soon after sprang up at southwest, and blew fresh, which rendered the sail boats of use, and at the same time made the passage from the island to the city, direct, easy and expeditious. Providence further interposed in favor of the retreating army, by sending a thick fog about two o'clock in the morning [August 30] which hung over Long-Island, while on the New York side it was clear.

Had it not been for the providential shifting of the wind, not more than half the army could possibly have crossed, and the remainder, with a number of general officers, and all the heavy ordnance at least, must inevitably have fallen into the enemy's hand. Had it not been for that heavenly messenger, the fog, to cover the first desertion of the lines, and the several proceedings of the Americans after day-break, they must have sustained considerable losses. The fog resembled a thick small mist, so that you could see but a little way before you. It was very unusual also to have a fog at that time of the year. My informer, a citizen of New-York, could not recollect his having known any at that season, within the space of twenty to thirty years.

The "heavenly messenger" was also described by others. "At this very time a very dense fog began to arise, and it seemed to settle in a peculiar manner over both encampments," wrote Major Tallmadge. "I recollect this peculiar occurrence perfectly well; and so dense was the atmosphere that I could scarsely discern a man at six yards' distance." Another soldier spoke in the same vein: "Providentially for us, a great fog arose, which prevented the enemy seeing our retreat from their works which were not more than a musket shot from us."

Washington had extricated his forces from one trap, but would soon find himself facing another. Philip Fithian took notice of the situation: "the winds are northerly & have been since they (the British) came on shore, but the huge ships beat up a little nearer every tide, & we hourly expect them before the town." When southerly or westerly wind flow ultimately replaced the long-continued northerly circulation, the British fleet gained the desired weather gauge and was able to navigate at will the waters surrounding the long narrow profile of Manhattan Island. Only one escape route from the island then remained for the Americans — over the Harlem River at Kingsbridge to the mainland in south Bronx.

Bennington: Muddy Victory

Every native of the Green Mountains is supposed to know (but seldom acknowledge) the fact that the Battle of Bennington was not fought on Vermont soil, but just over the border in New York, the territory of the hated "Yorkers" who laid claim to the area east to the Connecticut River in pre-Revolutionary times.

The British General John Burgoyne, in moving south down the Champlain and Hudson lowlands, needed horses and wagons to cover the portage between the two waterways. In mid-August of 1777, he sent a detachment of Hessian troops on a foraging expedition eastward from his base. Their progress was delayed by rains and high waters and by an excess of cumbersome equipment. They finally came into contact with the Americans west of Bennington and prepared for a pitched battle. But the intervention of an overnight rainstorm worked to the benefit of the rebels under General John Stark by delaying the start of the conflict for twenty-four hours and allowing time for additional units of militia to swell their numbers. Furthermore, the overnight deluges made difficult the preparation of earthwork defenses by the Germans. The hard rain washed down the dirt walls, filled the trenches with water, and made the troops miserable in their exposed position on a hillside.

Muddy roads and high water also delayed a relief column of German reinforcements. In the words of their commander,

> The crossing of the Battenkill [River] consumed considerable time, for the men had to wade through the water. The great number of hills, the bottomless roads, and a severe and continuous rain, made the march so tedious that I could scarcely make one half of an English mile an hour.

As a result of the slow progress of this relief column, General Stark's men were able to deal with the two enemy forces separately: they surrounded and captured the first column during the afternoon of August 16, 1777, and then, when darkness was falling, halted and then put to rout the second. The coming of darkness brought an end to the killing. "Had day lasted an hour longer," declared Stark, "we would have taken the whole body of them."

Burgoyne learned a bitter lesson. As he later wrote, "The New Hampshire Grants [Vermont] . . . a country unpeopled and almost unknown in the last war, now abounds with the most active and rebellious race on the continent, and hangs like a gathering storm upon my left." It would be another stroke from this storm country that would seal the fate of the British at Saratoga two months later.

Trenton: The Fight Against Ice and Sleet

The Crossing

Washington's plan called for three simultaneous crossings of the Delaware River on the early morning of December 26, 1776. The main force of some 2400 Continentals under his command would employ McKonkey's Ferry (now Washington Crossing) and form the northern spearhead. Another crossing in strength was to be made about 20 miles to the south at Bristol Ferry by General Cadwalader and a force of about 2000 Pennsylvania militia plus a New Hampshire brigade. In the center, at Trenton Ferry, about 600 Pennsylvania militia under General Ewing were to cross and hold a blocking position just south of Trenton.

The southern group attempted to cross at the appointed place above Bristol, "but the river was so full of ice that it was impossible to pass," Cadwalader wrote Washington. Another try was made six miles farther downstream at Dunk's Ferry at Pennsauken Creek. This was partially successful in landing some infantry.

Sergeant William Young's experience was much the same: "On account of the ice on the Jersey shore they could not land the Great Guns. Crossed back again, it came on to snow and rain. Wind E.N.E. Very Cold. Our men came home very cold and wet." General Cadwalader explained his failure to land the artillery: "It was impossible the ice being too thick." Without gun support, he made the controversial decision to recall the forward infantry back to the Pennsylvania side.

General Ewing, in the central position at Trenton Ferry, failed to land a man on the east bank; Washington commented on this on December 27: "The quantity of ice was so great that, though he did everything in his power to effect it, he could not get over." The William Fadden map of 1777 also stated that Ewing "could not get over on account of the quantity of ice."

Normally, ice conditions are worse above the falls of the Delaware than below. McKonkey's Ferry lies some 7.5 miles upstream from the falls just below Trenton. Washington, unlike the forces farther downstream, had the use of large boats employed in the transport of farm produce and iron goods from upstream. Known locally as Durham boats after their maker, they were said to measure from 40 to 80 feet in length, have a beam of 8 feet, and to draw only from 24 to 36 inches when fully loaded. They served as improvised ice breakers, since their weight plowed through the ice floes in midstream and their curved prows rode over and crushed the shore ice. Washington, soon after the crossing, wrote of "breaking a passage thro' the ice." The Durhams were manned by experienced boatmen from Gloucester, Massachusetts.

March and Attack

The weather scene at the northern crossing and on the march was described by a correspondent of the *Freeman's Journal* of Philadelphia: "About eleven o'clock at night it began snowing, and continued so until daybreak when a most violent northeast storm came on, of snow, rain, and hail together." All reports mentioned a mixed precipitation as the mercury hovered close to 32°F (0°C). One of Washington's staff wrote of conditions about 3:00 A.M.: "The storm is changing to sleet and cuts like a knife."

The periods of snow and ice pellets in the early morning and the near-freezing temperatures all day created extremely hazardous conditions for marching. "The night was sleety, and the road so slippery that it was daylight when we were two miles from Trenton," wrote a New Jersey rifleman. Fortunately, the northeast wind was at an angle to the backs of the men marching toward the southeast, so they were spared the discomfort of facing the stinging needles of ice and the cold drops of rain. For the defenders of Trenton, the wind blowing full in their faces cut down their vision and ability to discern the enemy.

A favorable effect of the storm and the attending delays in the crossing and on the march became evident when the advance columns encountered no patrols in approaching the first Hessian outposts shortly before 8:00 A.M. in the growing daylight. Though the Hessians exhibited good military discipline, with picket outposts maintained during the night on the principal roads north of town, the night pickets had been withdrawn and the day patrols had not yet gone out. The raging storm, no doubt, reduced their vigilance during the crucial hour after dawn.

The Americans were put at a disadvantage, however, by the wet conditions prevailing. Their flashpans and powder became wet from the driving precipitation, causing many muskets to misfire.

Momentarily surprised, the Hessian duty troops poured out of their barracks at the first shooting, and soon the 1500-man force attempted to form an organized resistance. But the Americans in the first rush succeeded in wheeling their guns into position to command the main streets, then several columns in a mad charge drove into the center of town despite their inability to fire their guns properly. The lively battle ended in less than an hour, with most of the enemy surrounded in the southern end of the town. About one third of the Hessian garrison succeeded in escaping southward, but the remainder were either killed, wounded, or captured. Only two Americans were reported to have lost their lives, and both were victims of the weather, having fallen exhausted on the march and frozen to death.

Princeton: A Providential Freeze

Soon after the news of the defeat at Trenton was received at headquarters, the British sought to retaliate. Under the energetic leadership of Lieutenant-General Cornwallis, they advanced from New Brunswick, through Princeton, to Maidenhead (now Lawrenceville) on January 1–2, 1777. The early January thaw prevailing on New Year's Day melted the snow and brought the frost out of the ground. The hard rain during the night added to the softness of the ground surface, turning the erstwhile hard roads into bottomless tracks and making the movement of men and vehicles most difficult. "It rained when we set out; on account of the thaw the road was muddy and very deep," observed William Young, and that night he complained, "I was a great deal fatigued on account of the deepness of the road." Marching up from Crosswicks at the time, Lieutenant Peale found the roads "very muddy, almost over our shoe tops."

The Americans had once again crossed the river on the twenty-ninth and thirtieth in order to exploit their victory more fully. They took positions in Trenton with advance parties to the northeast as far as the southern section of Maidenhead, where the British came upon them about midmorning on January 2. That afternoon the Continentals were forced back by enemy pressure to prepared defensive positions along the south bank of the narrow Assunpink Creek at the southern end of Trenton. The British made some efforts to cross the little stream late in the afternoon, but met with stout opposition. As daylight waned, Cornwallis believed he had Washington trapped and decided to wait till next morning to launch a full-scale attack and "bag the fox."

Washington now found his command in a desperate position: outnumbered — a well-equipped and disciplined enemy in front — confined to a small maneuver area by muddy roads on his right flank — an ice-clogged river on his left discouraging a recrossing — and a barren hinterland in his rear. A council of war was held late on the second from which an audacious plan emerged. "A providential change in the weather" seemed to be in the making.

A cold front had moved through the area early on the second, bringing an end to the rain, clearing the skies, and causing a shift of the wind. The day was described by the Philadelphia observer as fair and windy in the morning, and cloudy and still windy in the afternoon, a typical post-frontal day in wintertime. Washington noticed the wind holding to the northwest all day in the Trenton area and the temperature not rising during the daylight hours, as a steady advection of cold air nullified the warming effects of the sun. The Philadelphia thermometer read 39°F (3.9°C) at both morning and afternoon observations. From his past weather experiences at Mount Vernon, Washington realized that this meant a freeze that night, a freeze that would harden the roads into a tractable surface and permit the army, now immobilized by General Mud, to move.

While seemingly preparing to hold the defensive position overnight, with huge fires blazing and much moving about, Washington's entire force packed their equipment and supplies after midnight in preparation for a daring end run around the left flank of the British camped across the creek. A secondary road recently hewn-out provided an escape route. It ran east-north-eastward three or four miles to the southeast of the main Trenton–Princeton road, now in British hands, through portions of the present communities of Mercerville, Quaker Bridge, Clarksville, and Port Mercer, then converged to join the main Trenton post road at the Stony Brook bridge just south of Princeton, following for most of the way parts of present routes 535 and 533.

After sunset, the area thermometers dropped below the freezing point, as Washington had anticipated, to 31.5°F (−0.3°C) at 9:00 P.M. When the army began to move, about 1:00 A.M. on the third, it was even lower, and the desired freeze had taken place, as Captain Rodney noted, "at two o'clock this morning, the ground having been frozen by a keen N. West wind." The new road now presented a tractable surface. Captain Olney commented: "The roads which the day before had been mud, snow and water, were congealed now and had become hard as pavement and solid."

Washington hoped to attack Princeton before dawn and secure the advantage of surprise, but "the sun rose as we passed over Stony Brook," noticed Captain Rodney. This was less than two miles from their objective. Sunrise came at 7:23 A.M. according to the *Burlington Almanac for 1777*.

"The morning was bright, serene, and extremely cold, with a hoar frost which bespangled every object," recalled Lieutenant Wilkinson. It was a typical anticyclonic morning in wintertime. The first rays of the sun found Washington and his troops well in the rear of the main body of Cornwallis's force, still looking forward to the decisive frontal assault on the erstwhile American position south of Trenton. The advance guard of the forward detachment of Americans, when approaching Princeton from the southwest, came into contact with an equally small column of

British regulars moving on the post road toward Trenton. A clash took place between Stony Brook and Frog Hollow, astride present Mercer Road, southwest of Nassau Hall. The Philadelphia thermometer at this hour read 21°F (−6°C), and it may have been 2 or 3 degrees lower in the Princeton area, so the ground was solid — the open fields as well as the roads were suitable for military operations.

The Battle of Princeton was a hotly fought, but brief affair, the entire fire-fight lasting only about 45 minutes from start to finish. In the first phase, the disciplined British regulars routed the American left wing, composed mainly of militia, with a vigorous bayonet employment; but then Washington led the main body of seasoned Continentals in a victorious charge against the outnumbered enemy.

The successful outcome of the clash cleared the way for Washington to move across Cornwallis's main supply route and to threaten to occupy his main base at New Brunswick. Such a prize, however, was beyond the ability of the leg-weary Americans. Washington wisely selected his secondary objective: the strategic Watchung range of hills in north-central Jersey, a safe lair from immediate pursuit and a vantage point for keeping watch on the British.

The 15-hour duration of the freeze, from about 10:00 P.M. on the second to about 1:00 P.M. on the third, enabled Washington to disengage from a muddy confrontation with Cornwallis before Assunpink Creek, march 18 miles over tractable roads, maneuver to defeat a British rear guard before Princeton on a solid surface battlefield, collect booty in Princeton, and escape northward along the Millstone River before an afternoon thaw set in, turning the roads to mud again and making movement difficult.

Saratoga: A Dolorous March

In early October of 1777, General John Burgoyne was in a fix. Two attempts made at Bemis Heights on September 19 and October 7 to smash the American left wing and open the way to Albany had failed. He had hoped to join forces with General Clinton, who he thought was moving up the Hudson River from New York City. It was now early autumn, the time of year when nights begin to grow uncomfortably cool, the first frosts whiten the grass and fields in the morning, and the foliage takes on a golden hue under brilliant sunshine by day. Although it was still Indian Summer weather, winter was not far off.

On September 19 the fog had not lifted until 10:00 A.M., delaying the start of the battle until almost noon and perhaps preventing a decisive conclusion before darkness fell. The next morning was foggy, too, and the idle morning hours gave Burgoyne time for second thoughts about renewing the battle, since he had suffered sizable casualties. The next fighting, on October 7, did not seem to have any complicating weather factor, but the late start of the battle again prevented a decision just as the rebels were about to break into the British redoubts.

The final phase of the campaign commenced on October 8, when Burgoyne began a retreat northward toward Canada, now having despaired of succor from the south. With no meteorologist on his staff to advise him, he set out on a warm evening with an ominous southerly wind blowing. Soon it began to rain.

> The progress was slow beyond belief, not more than a mile an hour. It was a dolorous march. Rain was falling heavily. The road, bad enough before, was a bog. The tired men could hardly drag their feet out of the mud. The wagons stuck fast and were unable to go on. The tents and baggage were left behind,

wrote General von Riedesel, commander of the German mercenaries.

The Americans pursued the retreating force closely. On the eleventh, a near-disaster occurred when an American column, advancing through another morning fog and thinking they were closing on the British rear guard, almost marched into the enemy's main position. If a general engagement had developed under these circumstances, it might have permitted the British and Germans to withdraw unmolested. Finally, on the next day, New England militiamen under General John Stark succeeded in crossing the Hudson River north of Burgoyne's camp at Saratoga, effectively blocking the only escape route.

John Trumbull's painting of The Surrender of General Burgoyne at Saratoga. *Photo no. 16-AD-8 in the National Archives.*

Burgoyne asked for terms on the thirteenth, and the capitulation took place on the seventeenth. A factor in his decision may have been the advanced state of the season, since frosty nights and cold mornings already gave a hint of what was to come. The first snow fell on October 21, and a hard freeze followed. Another snow mixed with sleet and rain fell on the twenty-eighth. Winter had come.

On to the Brandywine: Sweltering

Again, Sir William Howe kept General Washington guessing as to his destination when a large British fleet cleared Sandy Hook and entered the Atlantic Ocean on July 22, 1777. After a rendezvous off the Delaware Capes on the thirty-first, Howe decided to take the long route to Philadelphia by sailing south around Cape Charles in Virginia and then north to the head of Chesapeake Bay. This would place his troops about 45 road miles southwest of the American capital at Philadelphia.

The troops were to spend six to seven weeks aboard their transports in the hottest time of the year. Major Carl Baurmeister summarized conditions aboard the ships: "During most of the voyage we had contrary winds and intense heat, which was accompanied almost daily by terrific thunderstorms, causing much suffering among men and horses and damage to the masts and sails."

The landing was made in Maryland on August 25 close to present Elkton where the Elk River empties into the bay. A strong southerly wind was blowing, the harbinger of a heavy rainstorm that commenced soon after the landings and continued into the next day. The wet conditions caused a postponement of the British march order because troops and equipment were soaked.

The skies finally cleared early on August 28 and the following day was "extremely fine." With the roads drying out, the British commenced their advance in the direction of Philadelphia. Meanwhile Washington, having hastened southward from New Jersey, collected a force to place between the British and the American capital. After a minor skirmish at Cooch's Bridge on September 3, the British encountered the main American army a week later on the Philadelphia road where it crosses Brandywine Creek at Chadds Ford close to the present Delaware-Pennsylvania border.

A little rain fell on the night of September 10–11, but the overhead skies cleared by 7:00 A.M., when the Philadelphia thermometer stood at 67°F (19°C). During the day there were both sunshine and clouds. Christopher Marshall at Lancaster described the morning of the eleventh as "cloudy, yet pleasant, cleared up warm." Along the Valley of the Brandywine it was "a very foggy morning and so dark we could not see a hundred yards before us," in the words of British Lieutenant Martin Hunter.

Howe decided on a strategy that would outflank the American right wing by making a long march north, then east. Its start was concealed by the fog, but as skies cleared, the day "ripened to a noon of blazing sunshine and sweltering heat." The route of the troops was described as "seventeen miles of very dusty road."

Since the early part of the British maneuver had been hidden by fog, several hours passed before Washington realized the purpose and strength of the British tactic and tardily sought to counter it. When the troops became locked in combat during the middle and late afternoon, no weather deterrents were present except the heat. Unfortunately, no reading was taken on the Philadelphia thermometer that afternoon.

Washington was able to extricate his troops in a semiorderly fashion as dusk approached, since the leg-weary British were not disposed to press their advantage farther. Once again the American army suffered casualties and gave ground, but the army remained an entity to continue the fight for independence.

Chester Valley: Deluge

Washington next sought to maneuver his army to the west and north of Philadelphia to protect the fords of the upper Schuylkill River and to prevent the enemy from turning his right wing and pinning him against the Delaware River. The British came on the American defense line along the South Valley Hills of Chester Valley near Frazer, a short distance west of Paoli in Pennsylvania.

Two preliminary skirmishes occurred and a major battle loomed, but the weather elements intervened when a hitherto intermittent light rain became a deluge. "An Equinoxial gale at N.E. accompanied by incessant rains" swept the area as a northeast storm roared up the Atlantic seaboard. The storm ended all possibilities of an engagement. Some recent historians have labeled the affair "The Battle of the Clouds," though this hardly seems an accurate descriptive from a meteorological point of view.

The significance of the storm lies in its preventing what might have been a decisive battle, with the Americans holding an unfavorable position.

Germantown: Foggy Misfortunes

Fog on the morning of October 4, 1777, proved an influential factor in the outcome of the Battle of Germantown. In his post-battle report Washington assessed its influence:

> By concealing from us the true situation the fog obliged us to act with more caution and less expedition than we could have wished, and gave the enemy time to recover from the effects of our first impression; and what was still more unfortunate, it served to keep our different parties in ignorance of each other's movements, and hindering their acting in concert. It also occasioned them to mistake one another for the enemy, which I believe, more than anything else, contributed to the misfortune which ensued.

The Philadelphia area was under the influence of an anticyclonic regime that made for morning fogginess. Various estimates have been made of the existing visibility on the morning of the fourth, from 50 paces to 100 yards. The fog was soon thickened with the smoke of battle. "The fog together with the smoke made it almost dark as night," recalled General Anthony Wayne.

The critical part of the battle centered around the Chew House, in which a group of British Light Infantry had barricaded themselves, using it as a blockhouse directly in the middle of the American advance. In attacking this, two American columns fired on each other, adding to the general confusion of the fog-shrouded scene. The British were able to recover from the first impetuous American assault, re-form their lines, and blunt the force of the attack.

Valley Forge: Barefoot in the Snow

The American troops moved into the campsite at Valley Forge on December 19, 1777, a day described ominously by the Reverend Henry Melchior Muhlenberg, who lived nearby at Trappe, as having "stormy winds and piercing cold." The Philadelphia weather observer noted conditions as fair, temperature 30°F (−1°C) in early afternoon, and a northwest wind blowing. The record has been preserved in the library of the American Philosophical Society.

The temperature hovered near the freezing mark for the next several days, with "clouds and sunshine" until Christmas night, when a snowstorm brought 4 inches by dawn. Another snow came on the night of the twenty-seventh. Following this, the wind went into the northwest, bringing a cold wave. The Reverend Muhlenberg described the twenty-eighth: "the snow lies deep and a stormy northwest wind makes it piercingly cold." The thermometer in the city tumbled all day and night to reach a reading of 6°F (−14°C) on the morning of the thirtieth. It must have been close to zero in the countryside, away from the smoke pots of the city.

Captain John Montresor with the British in Philadelphia wrote on the thirtieth, "Wind at N.W., amazing cold. Schuylkill frozen over . . . Delaware full of ice . . . This day is looked upon as entering winter quarters." Next morning there were 5–6 inches of snow on the ground and the thermometer read 10°F (−12°C). The year ended with fair skies, but the temperature remained below freezing.

General Nathaniel Greene wrote of the condition of his men: "One half of our troops are without breeches, shoes and stockings, and some thousands without blankets." A quarter of the number were reported unfit for duty "because they are barefoot and otherwise naked." Fortunately, this was the severest spell of weather to be experienced during the winter ordeal at Valley Forge.

Generally speaking, the winter of 1777–78 was an in-and-out affair. There were two periods of relatively severe cold: at the end of December, with a temperature of 6°F (−14°C) at 8:00 A.M. and at the beginning of March with a reading of 8°F (−13°C). Two periods of moderate cold came in midwinter: January's lowest was 12°F (−11°C) and February's, 16°F (−9°C).

Only three snowstorms were of any duration, and these were of the moderate, not heavy, variety. On December 28, 4 inches fell at Philadelphia, with a greater amount indicated for the Valley Forge area; a "deep snow" came on February 8 that was washed away by a heavy rain on the tenth and eleventh; and on March 2 and 3, "enough for sleighing" fell in downtown Philadelphia, concluding the snow season. A diarist at the camp indicated as much as a "foot of snow" on the ground early in January, and Dr. Muhlenberg at Trappe thought the snow of February 8 "deeper now than we have had the whole winter."

In January the ground appears to have been snow-covered a good part of the time, and in February there seem to have been more days with the ground bare. Throughout the season, frequent thaws created muddy fields and caused atrocious conditions on the roads for travel by horseback and movement by wagon.

The Delaware River was closed to navigation by ice during the cold spell from December 30 to January 8, and again for a few hours on January 19. Floating ice was seen on January 23, February 9, and March 5. But no long-continued freeze-up took place such as blocked the river for weeks in the winters of 1776–77 and 1779–80. There were several distinct, warmer-than-normal periods — December 20–26, January 3–12, 18–22, 19–31, and February 25–27 — when the temperature remained above freezing all night in Philadelphia.

The winter began to break up on March 9–10 as overnight thaws became regular. The twelfth brought the first truly spring weather. These dates are close to the normal for the onset of favorable spring conditions in the Philadelphia area.

On the basis of cold statistics, the winter of 1777–78 in the latitude of Philadelphia was not a severe one. Of the seven winters during the period of the War of Independence when the contending armies were in the field, two (1779–80 and 1781–82) were severe, and two were notably mild (1778–79 and 1780–81); the others (1775–76, 1776–77, and 1777–78) were moderate.

If Valley Forge had occurred a year earlier there would have been more snow and about the same cold; and, if a year later, conditions would have been less harsh. Nonetheless, one needs only to read the contemporary testimony of those who endured the winter at Valley Forge to realize that spending even a very mild winter in the field without proper equipment, food, and shelter would constitute an ordeal for a human being in the best of health.

A Hurricane Prolongs the War: The "French Storm" off Rhode Island

The fine harbor surrounding Newport in Narragansett Bay held a position of strategic importance during the first half of the War of Independence. Although occupied by British forces in December 1776, its possession was not challenged until the summer of 1778, when troops under Major General John Sullivan and a powerful fleet under Admiral Comte d'Estaing, newly arrived in American waters, attempted to wrest control from the British army. Upon learning of the threat, the British fleet under Admiral Richard Howe sailed from New York to give battle.

The French were effectively bottled up within the bay by the southwest winds that normally prevail on the Rhode Island shore in summertime. But between seven and eight o'clock on the morning of August 10, 1778, a wind squall occurred as a cold front moved through the area. The wind shifted abruptly to north and the prevailing thick, pre-frontal weather cleared. This gave d'Estaing the "weather gauge," or windward position, that all naval men desire when engaging an enemy.

The French quickly cut their cables and sailed out into Rhode Island Sound with "the wind moderate at NE." The two fleets were approximately equal in strength, but Admiral Howe, who had the wind in a disadvantageous quarter, decided to retire southwestward in a delaying action in anticipation of a return of the normal wind flow. With barometric pressure rising to the north and falling to the south, the wind held in a northerly quarter, enabling the French to close to about five or six miles of the rear of the enemy by sundown. On the next day, August 11, "the wind hung eastward, blowing fresh." At noon, conditions were logged as "thick hazy weather with moderate ENE breeze."

D'Estaing sensed victory, believing he could overtake the enemy and bring him to action while maintaining the wind advantage. His forward ships were preparing to engage the British rear late in the afternoon, but the wind increased from fresh to strong and then to a gale, the seas ran high, and sails were furled. These were the first strong puffs of a hurricane that was churning northeastward along the Atlantic seaboard; "hard gales" and

The French warship Languedoc *dismasted during a hurricane but still fighting the British warship* Renown *off Rhode Island. Courtesy of the Library of Congress.*

"strong gales" were sweeping the area by evening; by midnight they mounted to hurricane force as the storm center passed perhaps twenty-five to fifty miles eastward.

It proved a memorable night for both fleets. Dawn found d'Estaing's flagship *Languedoc*, the most powerful fighting ship afloat, almost totally dismasted and her rudder gone. The British were scattered over a wide area, and later limped by ones or twos into New York harbor. The French rigged jury masts and made minor repairs on the high seas and then headed for Boston, where there were adequate ship repair facilities.

The opportunity for a decisive naval action vanished under the blasts of the hurricane. The situation had similarities to the events in Chesapeake Bay three years later at Yorktown. It is interesting to speculate on how the war might have been shortened had the French prevailed in August 1778 and won naval supremacy in American waters.

Monmouth: Hot Pursuit

The British occupation of Philadelphia consumed many months in time, an enormous amount of military energy, and the expenditure of vast sums for supply and transport, all without accomplishing anything significant toward quelling the rebellious colonies. Washington's army was still in being and French reinforcements of unknown strength were known to be on the way. The British in the northern colonies held only the seaports of Philadelphia, New York, and Newport.

Upon taking command in America in the spring of 1778, General Henry Clinton decided that Philadelphia was not worth the price and ordered its evacuation. Part of his forces sailed down Delaware Bay, around Cape May, and along the New Jersey coast to New York Harbor, while the bulk of the infantry was ferried across the Delaware River for a 75-mile march across the sandy plains of New Jersey to the southern shore of New York Bay. Philadelphia was completely evacuated by the British troops on June 18. The season was now summer with hot, humid weather prevailing.

Upon receiving definite intelligence as to the direction of the British movement, Washington at Valley Forge set his army in pursuit across central New Jersey on a northeast course that would intercept the enemy's route before they reached the safety of the Monmouth Hills, backing on the shores of New York Bay. The paths of the two armies converged on June 27, with the Americans concentrating at Englishtown, separated by only 3 or 4 miles from the British strung out on the march near Monmouth Court House (now Freehold). It had been a hot march and a hot pursuit. In the words of Private Joseph Martin with the American van at Englishtown: "It was uncommonly hot weather and we put up booths to protect us from the heat of the sun, which was almost unsufferable."

The Battle of Monmouth commenced on Sunday morning, June 28, with a move by General Henry Lee against the British rear guard and baggage train then passing northeast of Freehold. The American attack, lacking coordination and resolution, turned into a disorganized retreat when the British reacted vigorously to the threat. At this juncture, Washington came on the scene and directed some torrid words at Lee for his unaggressive behavior. The American line finally stabilized and the enemy advance halted. Heavy fighting took place all day until dusk, when the British force withdrew and rejoined the marching columns.

Broiling weather featured the day of the battle as it had the previous days of march; Private Martin again described the weather on the battle day:

> The sun shining full upon the field, the soil of which was sandy, the mouth of a heated oven seemed to me to be a trifle hotter than this ploughed field; it was almost impossible to breath ... The weather was almost too hot to live in, and the British troops in the orchard were forced by the heat to shelter themselves from it under trees ... I presume everyone has heard of the heat that day, but none can realize it that did not feel it. Fighting is hot work in cool weather, how much more so in such weather as it was on the twenty-eighth of June 1778.

In a post-battle summary, Washington cited the debilitating effect of the heat as the reason for the failure of the American troops to pursue the withdrawing enemy:

> The extreme heat of the weather, the fatigue of the men from their march through a deep, sandy country, almost entirely destitute of water, and the distance the enemy had gained by marching in the night, made a pursuit impracticable. It would have answered no valuable purpose, and would have been fatal to numbers of our men — several of whom had died the previous day with heat.

How hot was it, actually, on Monmouth Day? One can read accounts stating that the thermometer stood at 92°F, 96°F, and 97°F (33°C to 36°C). Historian Gordon ten years after the battle wrote that it was 96°F (36°C) at Monmouth and 112°F (44°C) at Philadelphia that day, but gave no reference for his statement. The most quoted figure was 96°F, and this appeared in an account of the battle in one of General Clinton's letters to his sisters, dated July 6. This was probably the reading of the thermometer at the British headquarters in New York City. No thermometer reading for the Monmouth area has been located. The Philadelphia thermometer record, which would have been a good checkpoint for the degree of heat at Monmouth, ended in May 1778, leading to the supposition that the regular weather observer was a Tory and evacuated the city with Clinton's departure. There was a thermometer at Princeton during this period, but only its July readings are known.

The heat and thirst took quite a toll. A total of 112 British and German soldiers were killed in battle and 62 died from the heat, according to the diary account of Major Carl L. Baurmeister. American historians have ascribed the deaths of at least 37 missing Continentals to the heat also.

It is meteorologically appropriate that the symbol of the Battle of Monmouth should be Molly Pitcher, the legendary heroine, who succored many of the fatigued troops near her husband's artillery battery with cool water from a nearby well.

The Battle of Monmouth marked the last of the large-scale battles in the North involving a substantial number of troops. Thereafter, it became a war of maneuver, with most of the action taking place in Virginia and the Carolinas.

The Hard Winter of 1779–80

"The most hard difficult winter . . . that ever was known by any Person living."
— *Diary of Samuel Lane of New Hampshire*

During one winter only in recorded American meteorological history have *all* the salt-water inlets, harbors, and sounds of the Atlantic coastal plain, from North Carolina northeastward, frozen over and remained closed to navigation for the period of a full month and more. This occurred during what has ever since been called "The Hard Winter of 1780," a crucial period during the war when General Washington's poorly housed, ill-clad, and undernourished American troops at Morristown in the north Jersey hills were keeping a watchful eye on the British army much more comfortably quartered in New York City some 20 miles distant.

After five years of frustrating military activity, the fortunes of the Continental Army sank to a low ebb as the winter solstice arrived. Desertions were frequent, new recruits did not arrive, and mutiny was in the air among some units. Little did the men suspect in this discouraging atmosphere that nature had in store for them the severest season in all American history. They were almost as unprepared physically to cope with its rigors as they had been two years previously at Valley Forge facing a winter of only moderate severity.

The Hard Winter of 1780 went through three phases:

1. Snowbound. There came a series of exceptionally heavy snowstorms attended by much drifting in late December 1779 and early January 1780. These accumulated to depths of 2–3 feet in northeastern Pennsylvania and northern New Jersey and to 4 feet over much of New England. Though the final snow fell on January 6–7, the deep cover remained undiminished during the cold weather of the next five weeks.

2. Deep Freeze. "Frigidissime" was the diary entry of David Schultze of Goshenhoppen near Philadelphia to describe the cold of January. David Rittenhouse's thermometer went above freezing on only one day in January, the thirtieth, and several diaries reveal that the eaves on houses dripped on only one day during the month, presumably the thirtieth. The thermometer at the British headquarters in New York City was read at −16°F (−27°C), or just one degree Fahrenheit lower than any other reading in the past 110 years of official records. Up in New England even lower temperatures were registered. At Hartford, a trusty thermometer dropped to −22°F (−30°C), and the average sunrise reading for January was 4°F (−16°C), indicating a lower monthly mean temperature than any other in the modern records for the Connecticut capital.

It was cold in Virginia, too, as related by Judge George Mason, a recent widower. He wrote his friend James Mercer on February 5: "This cold weather has set all the young folks to providing Bed fellows. I have signed two or three Licenses every Day since I have been at home. I wish I knew where to get a good one myself; for I find cold sheets extremely disagreeable." The Judge did not take a chance on another cold winter, for he married again on April 8, 1780.

The connubial urge caused by the cold weather led to a domestic tragedy of large proportions near Lancaster in Pennsylvania. Thomas Hughes related the sad details in his diary: "Feb. 5

— Forty people crossing the Susquehanna in sleighs — being on their return from a wedding — the ice broke, and six and thirty were drowned — amongst the unfortunates the new married couple."

3. Icebound. By mid-January, General Boreas had taken complete command of the military situation by blocking highways with snow and spreading a thick icy sheath over all navigable waterways. At Philadelphia, the Delaware was frozen solid from December 21 to March 4. Chesapeake Bay was crossed on foot over the ice from shore to shore, and down in North Carolina soldiers going north took a short cut across the ice of Albemarle Sound. It was at New York Harbor that the closing of the harbor and rivers was most crucial. For a period of five weeks following January 15, no supply ships could get up to Manhattan, so the troops had to forage for supplies. The heaviest cannon were hauled across the ice to Staten Island to seize and fortify possible sortie sites of the Americans. The British command feared that the ice might afford a highway for a quick nighttime descent on the city, so they took extra precautions. It was not until February 13 that the first thaw arrived, and not until the twenty-first that a small ship could pass through the Narrows and reach Manhattan Island from the sea.

One can compare the winter of 1779–80 with the two previous landmark winters of 1697–98 and 1740–41, particularly in the length of time that waterways remained solidly frozen. All seem to agree, everything considered, that the winter of 1779–80 topped both the previous contenders for the title. Ezra Stiles, along with the weatherwise father of Noah Webster, had lived through the rigors of 1740–41. They gave their nod to the more recent winter. Two other weather-watching correspondents in different areas, Ebenezer Hazard in Philadelphia and Jeremy Belknap in Boston and New Hampshire, added their affirmatives for 1780. In the South, Thomas Jefferson in Virginia and Colonel William Fleming in Kentucky thought there had been nothing similar to the Hard Winter of 1780 in the American experience.

General George Washington, who had maintained a daily weather diary at Morristown through the hardest part of the winter, lent his authority to the question. On March 18, he wrote to

General Lafayette: "The oldest people now living in this Country do not remember so hard a Winter as the one we are now emerging from. In a word, the severity of the frost exceeded anything of the kind that had ever been experienced in this climate before."

At the conclusion of the unusual season a committee of the American Philosophical Society was formed "to make and collect observations on the effects of the severe and long-continued cold of last winter."

Chesapeake Bay: Saved by the Anticyclone

The naval command of North American waters had been the greatest British asset since the outbreak of war. They were able to choose their point of attack, transport their forces at will, and achieve military superiority. This freedom of action had been challenged by the French fleet under Admiral Comte d'Estaing in the summer of 1778 with indecisive results. Not until the late summer of 1781 did the French again mount a serious challenge to the British naval supremacy. In late August, Admiral Comte de Grasse arrived from the West Indies with 24 ships of the line and took a position inside the entrance to Chesapeake Bay, between Cape Charles and Cape Henry.

On the morning of September 5, a British fleet under Rear Admiral Thomas Graves, sent to relieve Cornwallis at Yorktown, arrived off the capes and found access to the bay blocked by the French. A third naval unit was also heading for this rendezvous: a squadron of eight ships of the line under the command of Admiral de Barras coming from France via a stopover at Newport, Rhode Island. Its presence in American waters was known to both de Grasse and Graves, but its immediate whereabouts remained a question mark. If the two French fleets should join, they would exercise undisputed control of the waters they patrolled.

Mindful of the developing strategic picture, Admiral de Grasse decided to issue forth from his safe lair into the open Atlantic and give battle to the British, hoping to draw his enemy away from the entrance to the bay and permit de Barras to enter undetected.

The Atlantic waters had been churned by a severe storm during the opening days of September, but by the morning of September 5 anticyclonic conditions prevailed along the Virginia shore, with light winds out of the northeast and a small sea running; the weather was described as "moderate and fair" in the log of H.M.S. *London*. In bearing down from the northeast, Graves held the valuable wind advantage, or "wind gauge" as sailors call it. In the words of British naval historian William M. James, "The most favorable conditions that could be imagined for his attack were present." Instead of making an immediate frontal attack on the leading French ships, Graves chose to follow classical naval doctrine by forming a long battle line so that each ship might engage a French counterpart individually. To accomplish this, he had to wear about on an opposite tack, bearing east-southeast, and run roughly parallel to the line of French ships emerging from the bay. With the enemy to his lee about 3 miles, the British could choose the time of engaging while still holding the important advantage of the weather gauge.

The early and midafternoon hours were consumed in maneuvering the lines into battle order, the winds continuing light. Not until 4:30 P.M., with the sun already well down in the west, did Graves give the signal to attack. There followed a period of heavy fire, with most of the ships getting into action, though the rear British ships were too distant to engage effectively. The van ships of both fleets came into close contact and several were mauled severely. During the fighting the wind shifted about 45 degrees to the east-northeast and east and continued light. About 6:30 P.M., with sunset near and the wind dying, the battle lines fell off and the firing ceased. The engagement has been described as "two and a half hours of cannonading."

Though both commanders had determined to renew the engagement, the day of the sixth was spent in repair of damage, the British seeming to have suffered somewhat more than the French. Winds continued very light, rendering control of the cumbersome vessels difficult. The anticyclone dominating the area was drifting slowly eastward; at evening, wind flow was described as "feeble." Dawn of the seventh found the sea calm. Not until 10:30 A.M. did a light breeze spring up, and this came from the southeast, indicating the ridge line or spine of the anticyclone had moved past the meridian of the Virginia capes. The weather was still described as "fine and clear." The return flow on the backside of the anticyclone gave the French the wind advantage for the first time, but it was not of sufficient strength to exploit by maneuvering.

The morning of the eighth continued the light wind conditions, but late in the day "it came on to blow pretty fresh," with thunder and lightning enlivening the formerly placid scene. This marked the end of the stable weather conditions under anticyclonic controls and heralded the approach of a low-pressure system with active weather moving from the west. The rising winds whipped up rough seas, causing the damaged ships to labor heavily and ship water. H.M.S. *Terrible* was in the most critical shape, with water gaining hourly on her pumps; it was ultimately necessary to abandon ship and then destroy her with fire.

The morning of the ninth brought "obscure and foggy" conditions, with the wind again back in the northeast; the front of the storm system had passed eastward and another weather cycle commenced. This air flow gave the British the wind gauge again, but they seemed in no mood to take advantage and close to attack. That afternoon, after four days of inaction, de Grasse decided to return to Chesapeake Bay. With a favoring wind from the east-northeast, he pressed all sail bearing northwest. This was the last time the British saw their foe.

De Grasse came within sight of Cape Henry late on the morning of the tenth, and there lying within the protection of the bay was what he hoped to see: Admiral de Barras's squadron of eight ships of the line and ten transports carrying vital siege artillery and heavy equipment for investment operations against Yorktown. The most welcome reinforcements had skirted the battle scene and slipped into the bay while the British were held offshore by de Grasse. The safe arrival of the supply ships and their vital contents was to provide the key to the entire campaign.

So it turned out that the anticyclonic days from September 5 to 8 were probably the most important "nonstorm" period in American history. Now the combined French fleets made up a small armada holding area naval supremacy. Admiral Graves had no alternative but to return to New York, refit, and hope for the arrival of additional ships. The French transports landed their valuable cargo safely on the peninsula behind Yorktown. The fate of Cornwallis was sealed.

Siege of Yorktown: Won by a Line Squall

Early autumn in tidewater Virginia can produce different weather regimes that either favor or hinder outdoor activities. At one extreme, when a trough of low atmospheric pressure persists along the Atlantic seaboard, cyclonic storms — either coastal northeasters or, more rarely, tropical disturbances — cause unsettled conditions with rain and gales continuing for two or three days at a time, and these types may repeat their performance several times over a period of two or three weeks. Outdoor activities are restricted. On the other hand, September and October often find high pressure prevailing for days at a time with settled weather featuring clear skies, light winds, and a hazy atmosphere. Now known as "Indian Summer," these pleasant periods favor outdoor work and play.

The investment operations against General Cornwallis and his troops at Yorktown peninsula, continuing over the 22-day period from September 28 to October 19, 1780, enjoyed the more favorable of Virginia's autumnal weather types. There were only two noteworthy breaks in the atmospheric serenity, and both turned out to be advantageous to the Franco-American cause.

No meteorological records employing scientific instruments were known to have been maintained in the Chesapeake region during the autumn of 1781. The British invasion earlier in the year had halted the observations of the Reverend James Madison at Williamsburg. He later wrote Thomas Jefferson: "The British robbed me of my Therm and Bar." Furthermore, his prior records of daily thermometer and barometer readings were destroyed when his home was consumed by fire in the winter of 1781–82. Our source of weather information for the campaign comes from the many diaries and memoirs of the participants.

The investment of Yorktown followed the classical methods of siege warfare. The French were well schooled in these and had the required tools and weapons. The operation entailed digging parallels or deep trenches so that the attackers could approach closer and closer to the defensive redoubts of the enemy under cover of their own protective earthworks. The digging of the first parallel, scheduled to commence on the evening of October 6, required a dark, moonless night so the men at work would not be targets for the British gunners located only about a half-mile away.

The desired conditions began to unfold during the day of the sixth, when a southeast rainstorm set in. General David Cobb noted the day as "rainy at times." The following night was described by Colonel J. B. Wright as "dark and rainy, perfect for the purpose in view." Timothy Pickering recalled: "It was fortunately cloudy, and it rained gently; otherwise the moon (just passed the full) might have proved very injurious, by discovering us to the enemy." And Surgeon James Thacher struck a familiar theme: "We were favored by Providence with a night of extreme darkness." The first entrenchments were started that night.

In the final stages of the siege, an unforeseen weather event played a most decisive role in thwarting a British attempt to break out of their encirclement. On the night of October 16–17, in a bold effort to save them, Cornwallis decided to ferry his regulars to Gloucester Point on the north shore of York River, where a small perimeter was being held by Lieutenant Colonel Banastre Tarleton. Here the opposing allied force was thought to be vulnerable to a determined attack. If successful in breaking out, Cornwallis hoped to work his way northward through tidewater Virginia and join British troops operating out of New York City.

Cornwallis estimated that sixteen flatboats, each making three round trips, would be able to transport the core of his best troops across the York River estuary. Each trip would require about two hours for the boats to cross, unload, and return. Accordingly, the selected regiments assembled on the evening of the sixteenth. The first trip began about an hour before midnight and accomplished the crossing without incident. The boats returned, again loaded, and put off. "But at this critical moment," Cornwallis recalled in his campaign report, "the weather from being moderate and calm, changed to a most violent storm of wind and rain and drove all the boats, some of which had troops aboard, down the river." Other accounts mentioned thunder and lightning accompanying the storm. "It was," wrote Elias Dayton, "almost as severe a storm as I ever remember to have seen."

The direction of the wind must have been from the northwest, as the flatboats of the second crossing were blown downriver in a southeasterly direction. Some came ashore about a mile below Yorktown, but two were carried as far as 5 miles and captured by the French at the mouth of the York River.

Though meteorological details are lacking to make an exact judgment, the storm was probably the result of a line squall thunderstorm passing through the area. Line squalls often precede a cold front passage by several hours and can create even more turbulent conditions than the front itself. Tarleton recalled that the storm continued for about two hours, with the strong winds ceasing about 2:00 A.M. David Cobb mentioned cold conditions on the night of the eighteenth continuing through the twentieth, indicating a large polar air mass with colder-than-normal temperatures as the dynamic driving force behind the line squall and cold front.

The adverse turn of the weather completely disrupted the attempted breakout. With his best troops split in two groups by the river, Cornwallis found himself in an extremely vulnerable position and was forced to recall the men who had already crossed to Gloucester Point. "Thus expired the last hope of the British army," Colonel Tarleton commented.

Next day about midmorning a drummer boy appeared on the British parapet to roll out a call for a parley, and soon an officer carrying a white flag joined him. The American guns ceased firing and negotiations on terms of surrender commenced. The formal surrender took place on October 19; the weather must have been so normal that no one thought to mention it in his diary.

Upon receiving the news from America, Lord North, the King's chief minister, declared while pacing up and down his apartment at 10 Downing Street: "Oh God. It is all over. It is all over."

The War of 1812

The Dawn's Early Light:
Fort McHenry in Baltimore Harbor

> The rest of the night the three Americans paced the deck, scarcely daring to think what daylight might bring. Again and again they pulled out their watches, trying to gauge when the dawn would come. Five o'clock, the first light of day at last tinged the sky. Out came the spyglass, but it was still too dark to make anything out. At 5:50 it was officially sunrise, but there was no sun today. The rain clouds hung low, and patches of mist swirled across the water, still keeping the night's secret intact.
>
> But it was growing brighter all the time, and soon an easterly breeze sprang up, flecking the Patapsco and clearing the air. Once again Key raised his glass — and this time he saw it. Standing out against the dull gray of the clouds and hills was Major Armistead's American flag.
>
> — Walter Lord, *The Dawn's Early Light*

The War of 1812 was the most unusual of all the military exploits of the United States. Its beginnings were fuzzy without any single event leading to the declaration of war, and it occurred at a time when the British government had already moved to make concessions to the American demands since the nonintercourse and European System policies were causing unwonted economic distress in England. Then, at its conclusion, the major battle of the war at New Orleans was fought after the peace treaty had been signed in Europe. Both the Canadian and American capitals were burned without reason or military gain. Repeated inept leadership and unexpected military reverses on both sides led a recent historian of the bizarre affair to apply the label "The Poltroon's War."

We are unable to present any worthwhile weather analysis of many of the far-flung campaigns because weather records were not maintained either by the military participants or by the citizens in close proximity to the battle sites, nor are there any informative diaries of soldiers containing weather entries.

The Windy Burning of Washington

The nation's capital underwent a double ordeal on August 25, 1814, when man and nature combined to deal the city its greatest disasters ever suffered from either war or the elements, and these occurred simultaneously. While the public buildings were in flames from the British torch, a tornado smashed through the center of the city, doing major structural damage in the residential section and maiming many of the attacking force. More British soldiers were killed and injured by this stroke of nature than from all the firepower the American troops were able to muster in their ineffective defense of the capital.

Though crowded by military and political news, the Washington *National Intelligencer* gave space to the unusual meteorological event:

> On Thursday 25th last, while our devoted City was in possession of the Enemy, it was visited by a tremendous hurricane, which did great damage to the houses, blowing off the roofs of many, destroying chimneys, fences, &c. In some parts of the city every house was more or less injured. Much injury was doubtless done higher up in the country, where indeed it leveled an immense number of trees, uprooting them here and there, but more frequently twisting them off at the roots.

A British military historian, George R. Gleig, who was noted for his vivid writing, was on hand to describe the tornado's impact:

> Of the prodigious force of the wind, it is impossible for you to form any conception. Roofs of houses were torn off by it, and whisked into the air like sheets of paper; while the rain accompanying it, resembled the rushing of a mighty cataract, rather than the dropping of a shower. The darkness was as great as if the sun had long set, and the last remains of twilight had come on, occasionally relieved by vivid lightning streaming through it, which, together, with the noise of the wind and the thunder, the crash of falling buildings, and the tearing of roofs as they were stript from the walls, produced the most appalling effect I have ever, and probably ever shall, witness. This lasted for two hours without intermission; during which time, many of the houses spared by us, were blown down; and thirty of our men, besides some of the inhabi-

tants, buried beneath their ruins. Our column was as completely dispersed, as if it had received a total defeat; some of the men flying for shelter behind walls and buildings, and others falling flat on the ground, to prevent themselves being carried away by the tempest; nay, such was the violence of the wind, that two pieces of cannon which stood upon the eminence, were fairly lifted from the ground, and borne several yards to the rear.

Certainly historian Gleig was impressed by the force of an American tornado. The only part of his account that might raise a meteorologist's eyebrow was the mention of the storm's continuing for two hours. Most likely, this was a reference to the thunderstorm that bred the tornado. Though the latter's passage across the city would take only two to three minutes, a large thunderstorm with expanding cells aloft might hover over the city for as long as two hours.

That this was a tornado and not a fire whirlwind arising from the heat of the burning buildings is attested by the fact that tornadic damage was reported on the same afternoon in Loudoun County near Leesburg, Virginia, about 30 miles northwest of Washington, where two persons were injured and much destruction was done to forests.

Whether this was a single tornado moving southeast, or twin funnels, or a tornado swarm cannot be ascertained at this late date. Nevertheless, a powerful tornado with destructive winds did make a direct hit on the heart of Washington at this crucial hour in the nation's history.

New Orleans: Foggy Victory

The Battle of New Orleans on January 8, 1815, provided the principal American land victory in the War of 1812, but it came two weeks after the signing of the Treaty of Ghent that ended the war, and thus had nothing to do with its outcome or the terms. It was like a post-season football game after the National Championship had already been awarded. Nevertheless, it left a good taste in the American mouth, which had savored few triumphs. Accordingly, its anniversary is still celebrated as Jackson Day.

The principal weather element present was fog, which hindered the British from forming up their columns at an early hour for a sunrise attack. The British commander, Thomas Pakenham, commented: "While we were talking, the streaks of day light began to appear, although the morning was dull, close and heavy, with the clouds almost touching the ground. It is now too late."

The fog seems to have been an asset for each side. The American Lieutenant Colonel Thomas Mullins observed: "There was a thick fog on the River, and for 40 or 50 yards on the River, the enemy could not perceive our columns, consequently their firing was completely at random."

From the same perspective, J. H. Eaton wrote: "A thick fog that obscured the morning, enabled [the British] to approach within a short distance of our entrenchment, before they were discovered."

The wind shifted during the early stages of the engagement to the advantage of the Americans. "Quite a breeze from the northwest sprang up, and as the works ran about northeast and southwest, it drifted the fog toward the enemy," according to Buell Jackson.

General Jackson made no mention of the weather factor in his dispatches to Washington; apparently he attributed the victory to himself and not to God or the elements.

The long-ending War of 1812 finally ended in 1815.

The War Between the States

Americans have fretted a little because their "Grand Army" could not advance through mud that came up to the horses' shoulders, and in which even seven-league boots would have stuck, though they had been worn as deftly as Ariel could have worn them. They talked as if no such thing had ever before been known to stay the march of armies; whereas all military operations have, to a greater or lesser extent, depended for their issue upon the softening or the hardening of the earth, or upon the clearing or the clouding of the sky. The elements have fought against this or that conqueror, or would-be conqueror, as the stars in their courses fought against Sisera; and the Kishon is not the only river that has through its rise put an end to the hopes of a tyrant. The condition of rivers, which

must be owing to the condition of the weather, has often colored events for ages, perhaps forever.
— Charles C. Hazewell, *Atlantic Monthly, May 1862*

Weather Observing During the Civil War

The admirable system of volunteer weather observing stations developed in the decade of the 1850s by the Smithsonian Institution under the direction of Professor Joseph Henry suffered severely during the chaos attending the Civil War. Though many Northern records were faithfully maintained so that continuous records from 1850 and before now exist, in the South the disruption of ordinary pursuits caused a neglect of scientific activities, and the severance of communications prevented the gathering of any reports, even though the observations might have been faithfully maintained for a while.

The difficulties confronting a Southern observer were illustrated by Professor James S. Lull of Columbus, Mississippi, who, though able to continue his observations, was forced to reduce the breadth of his work on account of a lack of writing paper on which to make his entries. Fortunately, he did maintain an apparently complete precipitation record and part of his temperature register during the war years, and these later found their way to the Smithsonian files in Washington, D.C.

Henry Ravenel of Aiken, South Carolina — a well-known botanist, advocate of scientific farming, and experienced weatherman — in a July 1861 letter to the *Charleston Courier* urged Southern weather observers to continue their work and to hold their monthly sheets until the Confederate government would be in a position to receive them. How many did so is not known. Very few Southern records ever reached the archives in Washington, and these were mainly from the Border States and even these are mostly fragmentary.

As the Union pressure increased on the South and beachheads were seized along the Atlantic and Gulf coasts to tighten the blockade, the Army Medical Corps established military hospitals and the surgeons as part of their stated duties took up their routine of thrice-daily weather observations. The most useful of these was established on Hilton Head Island and at Beaufort, South Carolina, only a short distance from Port Royal and Savannah, to provide a daily check on weather conditions attending Sherman's March to the Sea in November and December 1864 and his subsequent drive northward through the Carolinas in February and March 1865.

The most important early operations in the West took place in west-central Tennessee, and for attendant weather conditions Professor William M. Stewart of Glenwood Cottage, located three miles south of Clarksville, Montgomery County, maintained a daily weather watch throughout the war years. His location was only 17 miles east of Fort Donelson, where the first major engagement of the war in the West took place, and about 42 miles northwest of Nashville, the key objective in the subsequent campaigns of 1863 and 1864. Stewart's fine records, which include full barometric and psychrometric readings, were nearly complete for the period from 1852 to 1871, providing not only daily data of the Civil War period, but also covering a sufficient time span to form normals and means of comparison. Thus, the war years in Tennessee may be put in their proper contemporary climatological setting.

When Grant moved south and the main action centered around the Confederate bastion of Vicksburg in Mississippi, the weather station of Professor Lull at Columbus in Lowndes County, Mississippi, proved useful. Professor Lull lived some 175 miles northeast of Vicksburg, but in wintertime a sampling of the general weather conditions at Columbus would provide similarities to those experienced by the troops opposing each other at Vicksburg. In 1863 and 1864 the Nashville area of central Tennessee remained the hub of activities. During this period Paul Tavel, a printer and bookseller, maintained an interesting diary and weather record at the Tennessee capital. His daily accounts of temperature and sky conditions provided an on-the-spot meteorological record for the battle days. His notes and comments on the affairs of the day in his native French language enliven the diary.

The few major battles and many minor skirmishes between

the Union Army of the Potomac and the Confederate Army of Virginia pivoted, for the most part, in a semicircle south, west, and northwest of Washington, sometimes within sight of the Capitol and seldom, except for the Peninsula Campaign in 1862 and the final phases in late 1864 and early 1865, more than 75 miles away. For the state of the atmosphere at Antietam, Fredericksburg, Gettysburg, and The Wilderness, the weather watch maintained at the U.S. Naval Observatory, then in the southwestern part of the city, can be profitably employed. The same air masses as prevailed at the battlefields were sampled six times per day at Washington, and these observations supply an approximation of temperature and precipitation conditions and give evidence as to the general atmospheric flow.

When the battle scene shifted to southeastern Virginia, Fort Monroe, in Union hands throughout the war, supplied a base for naval and army operations, and provided a vantage point to check conditions in this active theater of operations.

Because the battle between the ironclads *Merrimac* and *Monitor* took place within sight of Fort Monroe, conditions on the historic day of March 9, 1862, can be described with firsthand meteorological data. Also McClellan's Peninsula Campaign in the spring of 1862 and Grant's final drive on Richmond in 1864–65 employed Fort Monroe as a principal supply base, and the major battles of each took place within 75 miles of the fort.

The practice of weather forecasting at this time had not yet acquired the mathematical tools that would make it a scientific pursuit. Though work toward developing an understanding of the circulation of the atmosphere and the behavior of storm systems had been undertaken at the Smithsonian Institution and by individual professors at colleges and institutions, no regular system for issuing daily forecasts had been organized. Certainly, many of the blunders attending the military campaigns might have been prevented if some knowledge of coming weather conditions had been at hand. Also the introduction of aerial activity in captive balloons demanded a much more precise acquaintance with prevailing upper-air currents than was available.

One interesting note on weather forecasting appeared in the papers of Abraham Lincoln. It involved a letter to the President from Francis L. Capen, a "certified practical meteorologist & expert in computing the changes of the weather," who promised that "thousands of lives & millions of dollars may be saved by the application of Science to War."

Memorandum Concerning
Francis L. Capen's Weather Forecasts

April 28, 1863

It seems to me Mr. Capen knows nothing about the weather, in advance. He told me three days ago that it could not rain again till the 30th.of April or 1st.of May. It is raining now & has been for ten hours. I can not spare any more time to Mr. Capen.

A. Lincoln

The infamous "Mud March," near Falmouth, Virginia, January 21, 1863, as drawn by A. R. Wand for Harper's Weekly. *Courtesy of the Library of Congress.*

After the infamous Mud March in January 1863, the *New York Evening Post,* in an article titled "The Weather and the Army," called attention to the lack of meteorological guidance in the army command:

> Doubtless it is in military navigation that the science of meteorology might be applied to most advantage. By the aid of the telegraph, storms can be announced many hours before they appear above the curve of the earth, and vessels warned not to leave harbor. It is by no means improbable that the *Monitor* might have been saved by this precaution. Although the Smithsonian Institute makes meteorology a subject of special study, we are not aware of any effort to derive practical services from it to the army or navy.

Aeronauts in the Civil War

Upon the outbreak of the Civil War several of the country's balloonists offered their services and their equipment to the warring forces. Their efforts were largely unproductive at the outset because their balloons were not designed to be tethered in one location and the provision of an adequate supply of gas in the field proved troublesome. Balloons had been employed previously in warfare, but never on the scale they were to be used in the opening years of the Civil War.

A positive contribution was made by balloonist John La Mountain at Fort Monroe on Chesapeake Bay in July 1861 when, from an altitude of 1400 feet, he spotted two concealed Confederate camps and was able to direct artillery fire on them. He demonstrated the versatility of balloon flight by making an ascent to 2000 feet from the deck of the armed transport *Fanny* in the James River. La Mountain also pioneered in making reconnaissance flights over enemy territory. He employed the stratagem of drifting westward across the enemy lines when the wind at ground level favored this; then, after discharging ballast, he rose into the prevailing westerly current at higher altitudes, which carried him back to friendly lines. It speaks much for La Mountain's courage and knowledge of aerial currents plus good luck that he was able to repeat this feat several times without a mishap.

When a formal Union balloon corps was organized, its leader

Lowe's balloon, the Washington, *flies from the deck of the balloon boat* Custis *near Budd's Ferry, Maryland. Smithsonian Institution Photo No. 30913A.*

was Thaddeus Lowe, an aeronaut of considerable experience, who had secured the backing of Joseph Henry, director of the Smithsonian Institution, and also had the support of President Lincoln for the project. Lowe's first contribution followed the Union defeat at Bull Run in July 1861 when, on an ascent over Washington to a height of 3 miles, he perceived that there were no Confederate troops moving on the city, thus helping to calm an incipient panic that was seizing the populace of the almost defenseless national capital.

Lowe proved an innovator as well as an able balloonist and administrator, by designing a new type of balloon to meet military needs. He set fifty seamstresses to work in Washington on the construction of seven air bags of a new type; these went into service as the *Intrepid, Constitution, United States, Washington, Eagle, Excelsior,* and *Union.* Each balloon was equipped with a telegraph system for immediate transmission of intelligence to the ground. As an alternate, color-coded signals in the form of small paper balloons were employed for signaling, and the equivalent in color flares was provided for nighttime use.

Each balloon had a powerful oxy-hydrogen searchlight with an 18-inch reflector for use on night flights to spot enemy activ-

Lowe's balloon, the Eagle, *in a storm. Courtesy of the Library of Congress.*

ities. Portable hydrogen generators were constructed and placed on horse-drawn carriages for field use. Under Lowe's direction, a coal barge was rebuilt with a special launch deck for use as a mobile balloon base. This vessel, the *G. W. Parke Custis*, entered service on the Potomac River in November 1861, being the first ship to be reconstructed for specific use as an aircraft carrier.

Balloons were employed to good effect during the Peninsula Campaign in May and June 1862 at the battles of Fair Oaks and Gaines' Mill. The chief effect of their presence over Union lines was in forcing the enemy to spend much time and effort in concealing their camps, minimizing troop movement by day, and depriving the troops of the comfort of fire by night.

Despite their apparent value, the Union forces abandoned the use of balloons early in 1863. The reason seemed to have been not technical, but bureaucratic. Lowe was not a commissioned officer, and his mixed corps of civilians and military men proved perplexing to conventional officers who were responsible for the household duties of feeding and billeting the men and for deploying and servicing their equipment. Further, the conflict became one more of movement than of positional warfare, and the logistics of most situations would not permit the transportation and supply of a balloon facility.

The Confederates, too, experimented with the aerial vehicle. A gas bag was made in Charleston and carried to Richmond, where it was inflated at the city gas works. It was then conveyed to the battlefield for use in the defense of Richmond by being tethered to a locomotive of the York River Railroad. Later, it was flown from the deck of an armed tug on the James River until the ship ran afoul of a Union ironclad and was sunk, and the balloon was captured.

Later, another was assembled at Charleston to keep watch on the Union forces attempting to surround the city. The bolts of silk employed were of different colors and patterns, giving birth to the romantic legend that the ladies of the Confederacy had sacrificed their best dresses for the cause, hence the name "Silk Dress Balloon." The aerial effort had an untoward ending when the gasbag broke loose in a high wind, drifted over the Union lines, and was captured.

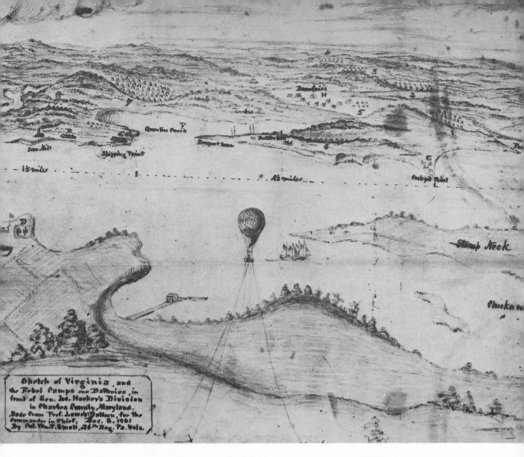

Balloon intelligence: Colonel Small reports to General Hooker on rebel camps and batteries in Virginia, December 8, 1861. Photo no. 94-X-2 in the National Archives.

First Battle of Bull Run: July 21, 1861

The first major engagement between the infantry forces of the North and South took place within 25 miles of Washington, along the small stream known as Bull Run, which drains southeast through Fairfax County into the Potomac River below Alexandria. With the war now three and a half months old, the North was eager to have a test of arms. Brigadier General Irwin Mc-Dowell, with 35,000 men, was ordered to advance into Virginia and seek battle with a Confederate force of 20,000 men under Brigadier General Pierre G. T. Beauregard, the victor of Fort Sumter and the current man of the hour in the South.

Moving out of Alexandria on July 16, the Federal force established a base at Centreville, facing the Confederates at Manassas Junction. Of great ultimate significance in the outcome was the arrival from the Shenandoah Valley on July 20 of 9000 additional men from Brigadier General Joseph E. Johnston's army, thus bolstering the Confederate force to make the manpower about even. It had been a relatively dry month, so the ground was solid and favorable for troop movement.

Both sides planned to attack on July 21, but McDowell took the initiative by starting his men on a flanking movement at 2:30 A.M., in bright moonlight, and struck later with a heavy artillery barrage at 5:15 A.M. He was successful in driving back the Confederate left and held this advantage until midafternoon. The arrival of more of Johnston's fresh units enabled the Confederates to counterattack, stem the Federal advance, and finally push back the enemy forces. The latter fell back in good order at first, but soon their retreat turned into a rout, approaching a panic as the military mixed with the assortment of spectators and Members of Congress who had come out from Washington to see the show.

In the early dawn hours of the twenty-first, at the beginning of the engagement, the sky was about half covered with clouds. The temperature on the Washington thermometer read 70°F (21°C), and the barometer had been rising since the passage of a weak cold front early on the twentieth. This had brought only a sprinkle of rain and introduced a northwesterly flow of modified polar air to the scene. The northwest wind blew across the battlefield all morning; it shifted to west during the first part of the afternoon and then went back to northwest. Sky coverage ranged from a low of two-tenths at 9:00 A.M. to a high of eight-tenths at 3:00 P.M., with both cumulus and cirrus clouds present. At the time of the decisive action in midafternoon the temperature reached a high of 84°F (29°C) at Washington. Since the Naval Observatory did not have a self-registering thermometer, the day's maximum might have been a little in excess of this figure. No rain was recorded on the twenty-first. It was an ideal day for outside activity. The main meteorological deterrent lay in the dryness of the ground, which caused great clouds of dust to rise. "The dust is denser than the smoke," reported the *Charleston Courier* in describing the battle scene, "for one long mile the whole valley is a boiling crater of dust and smoke."

Fort Henry and Fort Donelson: February 1862

As the year 1862 opened, the main strategy of the Federal forces in the West centered in a southward thrust through the Tennessee Valley. Penetration deep into the Upper South through the waterway would threaten from the rear the Confederate positions along the Mississippi River and would outflank the enemy's forces operating along the western slopes of the Appalachians in eastern Kentucky and Tennessee. The first obstacle to the advance of the combined Northern army and naval forces was Fort Henry, which guarded Tennessee River positions near the Kentucky-Tennessee border. The attack took place on February 6, 1862, with the gunboat flotilla pounding the unequal defenses into submission before the land forces had time to get into an attacking position. The Confederate defenders fled eastward 15 miles to Fort Donelson, on the Cumberland River.

The current weather played a part in weakening the defenses of the fort because the river, swollen by recent rains of 1.11 inches on the third and 1.37 inches on the night of the fifth–sixth, flooded the outer works of the fortification. These measurements were made by Professor William Stewart at Glenwood Cottage near Clarksville, Tennessee, situated some 35 miles east of the Fort Henry area. The precipitation "held up at 6:15 A.M." on the morning of the attack and no more fell during the day, although skies remained overcast.

Apparently, a cold front passed through the observation point at Clarksville about sunrise, having passed the battle area to the westward about an hour earlier. The temperature at 7:00 A.M. on February 6 read 58°F (14°C), with the wind blowing from the west-southwest. By early afternoon, the air flow went to north-northwest, the temperature fell to 56°F (13°C), and the barometer rose 0.15 inches (0.5 kPa) — all indications of the passage of a vigorous cold front.

The victorious Federal forces turned their attention eastward. Grant declared: "I shall take and destroy Fort Donelson on the eighth." But he failed to take counsel of "General Mud" and the swollen condition of the rivers and streams along the route. Cold air following the storm system on the sixth brought alternate nighttime freezes and daytime thaws, causing the roads to be rutted by night and bottomless by day. A very warm afternoon on the twelfth, with the mercury soaring to 69°F (21°C), caused many recruits to throw away their heavy overcoats and blankets, a common failing of new soldiers in their first campaigning when signs of "false spring" arrive.

It took Grant a week to move the distance of 15 miles overland to invest Fort Donelson. The naval attack opened on the afternoon of the fourteenth, but failed as this was a much tougher nut to crack than Fort Henry. A decided change in the weather situation took place soon after noon on the thirteenth when a sharp cold front introduced a frigid air mass, the coldest of the current winter season to penetrate the South. Temperatures tumbled from 58°F (14°C) at noon on the thirteenth to 11°F (−12°C) on the morning of the fourteenth, and snow fell to the depth of 2 inches. Grant said of the conditions: "The greatest suffering was from want of shelter. It was midwinter and during the siege we had rain and snow, thawing and freezing alternately."

The fourteenth proved a bitter day for troops operating either in the open or in the hastily constructed entrenchments. With northerly winds sweeping the area, the temperature rose only to 16°F (−9°C) at 2:00 P.M. on the fourteenth and to 18°F (−8°C) on the fifteenth. Clouds were scattered with lots of sunshine on the fourteenth, but overcast with some breaks in clouds prevailed on the fifteenth.

These extreme conditions, no doubt, greatly reduced the efficiency of attackers and defenders alike. They probably played a role in the decision of the Confederate command to evacuate the position as untenable in view of the dual onslaught from the enemy and nature.

Grant now occupied a favorable position to launch a spring campaign into western Tennessee and northern Mississippi with the objective of reducing the Southern strongholds along the eastern bank of the Mississippi River such as Memphis and Vicksburg.

The Monitor *was later lost in a storm off Cape Hatteras in December 1862. Rossiter Johnson,* Campfire and Battlefield *(New York: Bryan, Taylor Co., 1894).*

The <u>Merrimac</u> vs. the <u>Monitor</u>: March 9, 1862

A revolution in naval warfare came about on March 8, 1862, when the Confederate ironclad *Virginia*, formerly the U.S.S. *Merrimac*, appeared in Hampton Roads and began the destruction of the wooden-hulled Federal fleet engaged in blockading the ports of lower Chesapeake Bay. Satisfied with the damage and consternation caused that afternoon, the Confederate ram withdrew toward evening to await the completion of its task next morning. But at that very time the odd-shaped U.S.S. *Monitor*, described popularly as "a cheese box on a raft," steamed into the bay and took up a protective position off Fort Monroe, the Federal bastion commanding the lower Chesapeake throughout the war.

Next morning the unique ships engaged in a duel. The scene lay within the sight of Assistant Surgeon Sheldon, who had been maintaining the daily meteorological record required of all post medical officers. Thanks to his diligence, we have from within sight of the battle three observations of the weather conditions prevailing on the day of the historic engagement.

The fight took place during favorable anticyclonic conditions with calm seas and clear skies. A cold front passed through the area on the seventh attended by a light dusting of snow. The accompanying air mass was seasonally cool, but not cold, the atmosphere warming to a maximum of at least 44°F (7°C) on the eighth and 55°F (13°C) on the ninth, the battle day. Skies were fair on both days, and the light winds in the core of the anticyclone blew from the west, a direction usually ensuring good visibility on the bay. On-the-scene accounts mention light fog early in the morning of the day of the engagement, but this would appear to have been a thin morning fogginess caused by radiational cooling of the layer of air just above the warmer water surface, rather than an advective fog rolling in from the open ocean. The barometer peaked at 30.40 inches (103.0 kPa) on the morning of the ninth as the ridge line of the anticyclone passed over the area about the time that the fighting was under way.

Second Battle of Bull Run or Manassas Campaign: August 26–September 1, 1862

After a period of maneuvering in northern Virginia during August, the Union General John Pope found General "Stonewall" Jackson in strength on his northern flank and the enemy cavalry operating in his rear. Pope retreated from the Rappahannock River to concentrate his forces at Manassas Junction with a view of attacking Jackson. The latter withdrew westward to a stronger position in order to join forces with General Longstreet's army, moving eastward through the low defiles of the Bull Run mountains.

Pope attacked Jackson on August 29 and 30, was repulsed and thrown back to Henry House Hill, where his line held firmly through repeated Confederate attacks. A series of minor skirmishes took place next day, and the last scene of fighting occurred on September 1 at Chantilly, Virginia, in a heavy rain. The Federals held off the threat to their northern flank, and during the night drew back to the protection of the Washington area entrenchments, while Lee moved Jackson and Longstreet westward, where they could command the crossings of the Potomac above the capital. Pope's army was beaten, but not routed. Lee's army, though victorious, had failed to destroy Pope and open the way to Washington.

Weather conditions during the first half of August were favorable for campaigning, with much sunshine and only light rains falling. This regime was interrupted on August 28 by a day with early morning and evening showers that dropped 0.45 inch in the rain gauge at the Naval Observatory at Washington. A trough of low pressure was responsible for the cloudiness and rain on the twenty-eighth. But next morning, at the start of the first engagement, the low-pressure system passed eastward, the skies cleared, and the barometer was soon showing an upward trend. Winds were light until 9:00 A.M., when they picked up from the west and west-northwest to force 3 and blew briskly all day. The new air mass was slow in arriving, because it turned out to be a warm day with the thermometer mounting to 89°F (32°C) at 3:00 P.M. and the actual maximum might have been a degree or two higher. Fair weather cumulus covered seven-tenths of the sky at that hour.

The second day of fighting, Saturday, August 30, brought mostly cloudy conditions and light winds. The barometer reached a peak in its current cycle at 9:00 A.M., reading 30.23 inches (102.4 kPa). Cumulus clouds covered the entire sky at times during the day. Temperatures reflected the northerly flow and the presence of a new air mass by rising to only 74°F (23°C) at 3:00 P.M., a full 15 degrees below the previous day's reading. No rain fell during the day.

A shift of wind during the afternoon of the thirtieth heralded the approach of another weather system from the west. Beginning at 3:00 A.M., rain fell off and on during the day of the thirty-first in the amount of 0.35 inch, and on September 1, the concluding day of Second Bull Run, there "commenced an exceedingly heavy rain, with lightning and thunder, at 5:45 P.M." Amount of rain, 1.08 inches. Both days had a complete overcast during the daylight hours. Highest temperature readings on the thirty-first were 71°F (22°C); on the first, 79°F (26°C) — quite comfortable fighting weather for the area in late summer.

Battle of Antietam: September 17, 1862

A variety of reasons apparently led General Lee to invade the North in early September 1862. He crossed the Potomac River on September 4, and by the seventh had concentrated his relatively small army around Frederick in west-central Maryland, some 40 miles northwest of Washington. His overall plan was opportunistic and vague; perhaps he would strike northward toward Harrisburg to disrupt Federal communications and rail traffic, or advance eastward to threaten Washington and Baltimore from the rear.

General McClellan, with his usual lethargy, was slow in getting under way, and did not assemble his army in the Frederick area until after the thirteenth. The Union capture of the passes over South Mountain forced the Confederates to concentrate at the village of Sharpsburg, Maryland, with the Potomac on their southern flank and little Antietam Creek in their fore.

The weather during the middle days of September preceding the battle was generally favorable for the movement of troops under the anticyclonic conditions prevailing. On the day of the battle, the seventeenth, skies were completely overcast from sunrise to sundown with periods of light precipitation scattered over the area. Frederick reported light rain at 7:00 A.M. for a total fall of 0.15 inch. Sykesville had a "fine rain" ending at 4:00 P.M. for a total catch of 0.25 inch. The Naval Observatory at Washington reported rain at 2:45 A.M. and again at 3:00 P.M., amounting to 0.13 inch in all. A Georgetown observer mentioned "misting" at 7:00 A.M., and at Carlisle in Pennsylvania rain commenced at 5:00 A.M. and fell at intervals through the day. The rain seems to have been of a spotty, showery nature and not of sufficient intensity to interfere with troop movements or battle formations.

After a day's savage fighting on the seventeenth, Lee suffered such losses that he was forced to break off the engagement and retire across the Potomac on the night of the eighteenth to nineteenth. He was permitted to do so without molestation from McClellan's vastly superior force. The carnage had been atrocious. Total Confederate casualties — killed, wounded, captured, and missing — were about 13,700, or 35 percent of those engaged; Union losses were substantial, being placed at 12,350, or 18 percent.

Through the Eye of a Hurricane with C.S.S. Alabama

Hurricanes were noticeably absent on the Atlantic and Gulf coasts during the years of the war. None was reported to have crossed a coastline, but at sea the famous Confederate raider *Alabama* had the unusual experience of passing right through the central vortex or eye of a strong hurricane. Admiral Raphael Semmes in *Memoirs of Service Afloat During the War Between the States* described the passage through the eye of the storm on October 16, 1862, when the ship was in a position near 42°N, 60°W, or about 600 miles directly east of Cape Cod in the North Atlantic shipping lane. Let Semmes give his graphic description of this experience:

> The storm raged violently for two hours, the barometer settling all the while until it reached 28.64 [inches]. It then fell suddenly calm. Landsmen have heard of an "ominous" calm, but this calm seemed to us almost like the fiat of death. We knew at once that we were in the terrible vortex of a cyclone, from which so few mariners have ever escaped to tell the tale! . . . The scene was the most remarkable I had ever witnessed. The ship, which had been pressed over only a moment before by the fury of the gale, as described, had now righted, and the heavy storm staysail, which, notwithstanding its diminutive size, had required two stout tackles to confine it to the deck, was now, for want of wind to keep it steady, jerking those tackles about as though it would snap them in pieces as the ship rolled to and fro!
>
> The aspect of the heavens was appalling. The clouds were writhing and twisting, like so many huge serpents engaged in combat, and hung so low, in the thin air of the vortex, as almost to touch

Confederate warship Alabama. *Courtesy of the Library of Congress.*

our mast-heads. The best description I can give of the sea is that of a number of huge watery cones, — for the waves seemed now in the diminished pressure of the atmosphere in the vortex to jut up into the sky, and assume conical shapes, — that were dancing an infernal reel played by some necromancer. They were not running in any given direction, there being no longer any wind to drive them, but were jostling each other, like drunken men in a crowd, and threatening every moment to topple one upon the other.

With watch in hand I noticed the passage of the vortex. It was just thirty minutes in passing. The gale had left us with the wind from the southwest; the ship the moment she emerged from the vortex took the wind from the northwest. We could see it coming upon the waters. The disorderly seas were now no longer jostling each other; the infernal reel had ended, the cones had lowered their rebellious heads as they felt the renewed pressure of the atmosphere, and were being driven like so many obedient slaves before the raging blast. The tops of the waves were literally cut off by the force of the winds, and dashed hundreds of yards in blinding spray. The wind now struck us "butt end foremost," throwing the ship over in an instant, as before, and threatening to jerk the little storm sail from its bolt ropes. It was impossible to raise one's head above the rail, and difficult to breathe for a few seconds. We could do nothing but cower under the weather bulwarks, and hold on to the belaying pins, or whatever objects presented themselves, to prevent being dashed to leeward, or swept overboard. The gale raged, now, precisely as long as it [had] done before we entered the vortex, — two hours, — showing how accurately Nature had drawn her circle.

The Mud March: January 20–23, 1863

The most spectacular instance of weather controlling a military operation during the Civil War occurred in January 1863, when General Ambrose Burnside, smarting under his failure before Fredericksburg six weeks earlier, attempted to round Lee's left flank with a surprise, sideways march during the height of the Virginia winter.

The weather map was active, as it often is in mid-January. The major anticyclone of the winter season covered northern Virginia on the seventeenth and eighteenth with pressure cresting at 30.75 inches (104.1 kPa) at Washington on the morning of the eighteenth. The path of the center of the anticyclone lay to the north of Washington, probably across upper New York State and northern New England. This situation in wintertime often breeds a storm far to the south, in the Gulf of Mexico or along the South Atlantic coast. When the winds at Washington shifted into the northeast on the nineteenth, it confirmed that this development was taking place. With winds in the northeast, the barometer falling rapidly, and the skies clouding over, dawn of the twentieth presaged a major storm in the making. But there were no weather forecasting services in existence at that time to warn the commanders of what was in store for the next few hours.

Burnside commenced his complicated lateral move over an inadequate, primitive road system that became crowded far beyond its capacity. Colonel de Trobriand participated in the operation and wrote:

> On the 20th the division started at noon. The atmosphere was filled with moisture. The lowering heavens were without sunshine and without warmth. Over the roads still hard, through fields, and under forests, our long columns of infantry marched till night,

mingled with batteries of artillery, ammunition trains, and wagons carrying pontoons. . . . But, at the very time when we stacked arms, the fog, becoming more and more dense, turned to rain, which continued to fall cold, heavy, incessant.

The precipitation canopy of the storm advancing up the Atlantic seaboard reached Washington at 10:00 P.M. on the evening of the twentieth, and arrived in the battle area perhaps two or three hours earlier. Only 0.03 inch fell before midnight, but during the day of the twenty-first the Washington rain gauge caught the very heavy amount of 2.00 inches. Winds mounted to gale force during the day, commencing at northeast, then backing in midafternoon to north, when the barometer was close to its lowest point at a stated observation time, 29.75 inches (100.7 kPa). The weather at Washington that day was described as "squally, with heavy rain."

Back at the scene of the march, Colonel de Trobriand continued:

The rain lasted thirty hours without cessation. To understand the effect, one must have lived in Virginia through a winter. The roads are nothing but dirt roads. The mud is not simply on the surface, but penetrates the ground to a great depth. It appears as though the water, after passing through a first bed of clay, soaked into some kind of earth without any consistency. As soon as the hardened crust on the surface is softened, everything is buried in a sticky paste mixed with liquid mud, in which, with my own eyes, I have seen teams of mules buried. That was our condition on the 21st of January, 1863.

By evening of the twenty-first the entire army was bogged down in the mud. There was no alternative but to try to slosh back to the camps, which was done on the twenty-second and twenty-third. The Mud March had a much more significant influence than merely canceling Burnside's tactical move. Thereafter, no more major winter campaigns were attempted in the Virginia theater of war. General Grant, who had experience with a mixture of mud and winter at Forts Henry and Donelson, spent his first winter in command in the East in winter quarters at Culpeper, Virginia, and the final winter of 1864–65 in fixed position before the siege of Petersburg without field movement.

Chancellorsville: May 1–3, 1863

After the devastating defeat at Fredericksburg and the fiasco of the Mud March, the North looked for some action from the Army of the Potomac as soon as the coming of spring and weather conditions would permit. "Fighting Joe" Hooker, who gained a reputation as an aggressive division leader, had relieved General Burnside as commander. Accordingly, Hooker moved out of his camp site opposite Fredericksburg on April 27, 1863, crossed the fords of the Rappahannock River to the west, and took up a position at Chancellorsville, a crossroad site consisting of a single house and outbuildings in the midst of north-central Virginia's "Wilderness." The name was derived from the region's poor soil and impenetrable ground cover of thickets and small second-growth trees. Hooker intended to move around the extreme left of Lee's position in the entrenchments facing Fredericksburg and threaten the Confederate communications with Richmond.

Though numerically outnumbered on this front, Lee came out of his defensive stance in an attempt to check Hooker's advance. This he did. For reasons unexplained, the Federal forces failed to counterattack and went on the defensive in a perimeter around Chancellorsville. At this juncture, Lee tossed accepted military practices to the winds by splitting his troops and sending "Stonewall" Jackson on a daring flanking maneuver to strike Hooker's unprotected right flank. The blow delivered in force on the evening of May 2 caused the Federals to recoil toward the river into a defensive enclave protecting the same fords that they had so confidently crossed only four days before. Moderate to heavy rains fell on the fourth, fifth, and sixth, amounting to 2.35 inches. Finally, on the sixth, Hooker returned to the north bank of the Rappahannock; the crossing was made hazardous by the swollen stream. A thunderstorm on the morning of the sixth was credited with discouraging Lee from delivering a knockout blow while the withdrawal maneuver was in progress.

No rain fell on the battle days. Skies were partly cloudy on the first and second with never more than one-half coverage, but this increased to as much as eight-tenths on the third as a weather system approached from the southwest. On the first two nights

---

the moon was described by de Trobriand: "May 1 Firing ceased a little after dark. The moon rose calm and smiling, and nothing troubled the tranquility of the night. May 2 It was ten o'clock at night. The moon, high in the heavens, gave but an uncertain light through the vapors floating in the atmosphere." Temperatures on the first two critical days, May 1–2, ranged from morning lows of 53° to 55°F (12° to 13°C), and the afternoon readings on each day were 76°F (24°C) on the Naval Observatory thermometer at Washington.

The dry conditions existing on the first three days of the month enabled Lee and Jackson to plan and execute their daring maneuvers. Then the heavy showers on the next three days impeded the Confederate efforts to cut off the Union troops from their bridgehead, and the high water prevented them from taking advantage of other fords and getting in the rear of their enemy.

Precipitation conditions in a shower situation might vary considerably between Chancellorsville and Washington, but diary accounts told of "heavy" amounts at the battle scene. The Naval Observatory gauge at Washington registered the following catches: the fourth — 0.32 inch, the fifth — 1.10 inches, and the sixth — 0.93 inch. These were substantial amounts, sufficient to turn the red soil of the region into clinging mud.

Gettysburg: July 1–3, 1863

"Great battles of Gettysburg fought" was the simple entry of Michael Jacobs in his meteorological register for July 1–3, 1863. During almost twenty-five years at his post as professor of mathematics and science in Pennsylvania College, Jacobs had been noting the vagaries of the weather scene at Gettysburg for the Franklin Institute of Philadelphia and the Smithsonian Institution of Washington.

It was one of those incredible turns of fate that brought a horde of some 200,000 men, all bent on killing each other, from as far as the vast reaches of the Texas plains on the south and from the depths of the Maine woods on the north, to the tranquil valley of Gettysburg in south-central Pennsylvania, a remote pastoral retreat that Professor Jacobs loved so well. He had tramped every acre of its woods and fields in the company of students and friends on nature walks, during geological surveys, and for nights of sky-viewing to watch stars, meteors, and perhaps an aurora. More than anyone else he knew what nature had placed there for man's use and enjoyment. But now with his own eyes he was to witness his land desecrated by man's insanity.

The general weather map on the opening days of July 1863 presented a picture of stagnant conditions typical of summertime. A Bermuda High situation prevailed over the Eastern States, with anticyclonic strengthening on the first and second, to reach a pressure climax on the morning of the third and then remain about steady with small diurnal variations for the next twenty-four hours.

A study of weather station reports from the Great Lakes area and the Ohio Valley from July 1 to 5 indicated that no weather fronts were present, at least as far west as the Mississippi River. Furthermore, there were no significant wind-direction shifts or sharp wind-speed increases that would indicate anything but a very flat pressure gradient and the absence of any severe local storms.

All of the various weather elements observed by Professor Jacobs at his college weather station combined to indicate the transformation during the battle period of a modified polar air mass into one having tropical characteristics. The temperature, moderate for the season on the first two days of July, turned warm, but not excessively so, on the third. The sky remained overcast on the first, the clouds were broken on the second, covering about half the sky, and scattered over only a small portion of the sky on the third. Wind movement on all three days was light to gentle, at force 1 or 2, the two lowest on the scale employed for estimating the wind speed.

For military operations, the battlefield was solid and trackable for artillery and cavalry on all three days. Though June had been a drier-than-normal month, a heavy rainfall of 1.28 inches fell on June 25–27 and a lighter amount of 0.15 inch on June 29–30. No rain fell at Gettysburg during the three days of actual fighting.

Temperatures on the first and second days remained below

the seasonal normals with the 2:00 P.M. readings at 74°F (23°C) on the first and at 81°F (27°C) on the second. The thermometer on the third and crucial day rose from 74°F (23°C) at 7:00 A.M. to a reading of 87°F (31°C) at 2:00 P.M. At the "High Tide of the Confederacy" about 3:30 P.M., the mercury must have stood at or just below 90°F (32°C). Professor Jacobs did not have a self-registering thermometer that would have given the exact figure. Out on the field of battle, of course, sun temperatures through the scattered clouds would have the effect of seemingly higher readings. Incidentally, the reading of 87°F on the third was the highest observed at Gettysburg for the entire month and equal to the maximum for the whole summer.

The meteorological elements that proved most trying to the fighting men on the third were the relatively high moisture content of the air and the lack of wind movement. With tropical air now present and humidities running high, the comfort index soared to uncomfortable levels, and perspiration would have been the rule of the afternoon, even without the presence of an enemy army only a few yards away. If pictorial representations of the clothing worn by the combatants are realistic, the heavy blouses and trousers, suitable for winter campaigning, must have greatly increased the discomfort of the ordinary soldiers.

Another atmospheric element caused trouble at times. With the very light winds prevailing under anticyclonic conditions, the air toward the conclusion of the battles became very acrid. "By six o'clock the atmosphere had become filled with smoke, like a thick fog, and I could see only the flash of the artillery through this thick smoke," commented Private Henry R. Berkeley.

With the repulse of the last desperate charge, the fighting soon died down as the Confederates struggled back to their lines to await the expected Union counterattack. But none came, and silence fell on the battlefield. "The most profound calm reigned now, where a few hours before so furious a tempest had raged. The moon, with her smiling face, mounted up in the starry heavens, as at Chancellorsville," commented de Trobriand.

The weather elements, having been quiescent during the actual fighting, now assumed a more active role that was to complicate both Lee's withdrawal from the battlefield and General

Meade's attempts at pursuit. The stagnant anticyclonic conditions gave way late on the third to a showery period that continued intermittently until Lee reached the safety of Virginia on the thirteenth.

A light shower fell at Washington, D.C., on the afternoon of the third, and at Harrisburg, Pennsylvania, a "thundergust," commencing at 8:30 P.M. of the third and continuing until 6:00 P.M. of the fourth, dropped the sizable amount of 2.79 inches. Professor Jacobs's rain gauge showed a measurement of 1.39 inches on the fourth. If the fighting had extended another twenty-four hours, the heavy rainfall might have proved a decisive element on the battlefield.

Lee began his withdrawal from Gettysburg after dark on the fourth. "The night was very bad — thunder and lightning, torrents of rain," wrote Colonel Fremantle, a British correspondent, "the roads knee deep in mud and water." Lee's main force, making only one mile an hour, reached Hagerstown, Maryland, on the sixth to seventh, and Williamsport, on the Potomac River, on the eighth to ninth.

The southerly and southwesterly flow prevailing since July 1 gave way to a northeasterly flow on the night of the sixth to seventh, and a second heavy rainfall attending a general storm followed. The Harrisburg rain gauge caught 3.51 inches on the eighth while Professor Jacobs measured only 1.30 inches in his gauge at Gettysburg, differences typical of shower conditions. Greater amounts probably fell on the slopes of the mountains to the west of Lee's line of retreat.

The temporary pontoon bridge built by the Southerners across the Potomac had been partially destroyed by Union raiders. Lee, with his back to the stream now in high flood, found himself in a precarious position. The rains of the eighth raised the river to a near-record height. Even the city of Washington was isolated on the east by a washout of a key railroad bridge during these downpours. A strong defensive perimeter was thrown around the Confederate position between Williamsport and Falling River; so strong was the position judged to be that Union generals were hesitant to urge an attack.

No doubt the wretched condition of the roads slowed the

Northern army's endeavors to bring the necessary troops and artillery to the scene. Lee quickly reconstructed the damaged pontoon bridge to carry his artillery and supplies across, but he had to wait for the river to drop low enough to enable the infantry to ford the stream on foot, since the lone bridge did not have the capacity to concentrate the passage of so many troops in the short time that could be allotted to the maneuver. The Southerners began crossing the Potomac on the evening of the thirteenth and completed the dangerous operation by 1:00 P.M. the next day without interference. They were now safe for the present and still an army in being despite staggering losses.

Chattanooga and Lookout Mountain: The Battle Above the Clouds

The preliminary to the campaign to control the vital Chattanooga communications area took place at Chickamauga Creek on September 19–20, 1863. With the approach of the autumnal equinox, the first invasion of polar air from Canada reached eastern Tennessee. Frost was noticed for the first time on the morning of the nineteenth. Paul Tavel's thermometer at Nashville dropped to 38°F (3°C) on the nineteenth and to 34°F (1°C) on the twentieth. Some fog formed early in the mornings, but skies soon became clear as the sun rose in the heavens. Weather played little or no part in the Union success that followed.

The main battles for control of the railroad and waterway hub took place in late October and November 1863, with current weather conditions playing a vital role in the action. The first important move came on the night of October 26, when a force of 1500 engineers moved downstream past Confederate positions under cover of a heavy fog to Brown's Ferry, where a pontoon bridge was constructed and a beachhead on the eastern shore of the Tennessee River secured. This permitted General Hooker's corps to close on the southern flank of the Confederate defense line around Chattanooga. A full attack on this position was delayed awaiting the arrival of General Sherman's divisions, which had been held up by the heavy rains of mid-November.

Fighting in the "Battle Above the Clouds" at Lookout Mountain. Courtesy of the Library of Congress.

Early on November 24, Hooker and Sherman attacked with the objective of clearing Lookout Mountain of the enemy. This was a commanding position with an elevation of 2126 feet just south of the city, dominating the great bend of the Tennessee River.

While the battle crescendo increased on the heights during the afternoon of November 24, the thickening cloud deck gradually shut out the daylight, darkening the landscape more and more. As the moist air aloft condensed into cloud, the base of the overcast crept lower and lower down the side of the mountain and obscured all that went on above. By early evening, however, with the passage of a cold front and a wind shift to west, the clouds commenced to break and moonlight later revealed the Federal troops in clear command of the dominating summit, which had been wrested from a numerically smaller Confederate force. The latter retired to a stronger defensive position on Missionary Ridge to the northeast.

The cold front that moved through the area during the late

afternoon was responsible for the lowered visibility and the descending cloud level that marked the final stages of the battle. The nearest meteorological observers to the battle scene were some distance to the northwest, fortunately in the direction from which the front was coming. Paul Tavel resided at Nashville some 160 miles away and Professor Stewart was located at Clarksville, Tennessee, about 190 miles distant. Both reported light rain from overcast skies at their early morning observations on the twenty-fourth. The wind at Clarksville was from the south-southwest, but by the early afternoon observation had shifted to the west, the skies cleared, and the barometer began to rise. A cold front moved through in the morning, probably soon after 8:30 A.M., when Stewart noticed a drizzle commencing.

If the front traveled at 30 miles per hour (48 km/h), a reasonable speed, it would require about six hours to reach the Chattanooga area from Clarksville. Almost any near-normal rate of advance would put the cold front in the battle area during mid- or late afternoon, when the fighting was in progress. The arrival and departure of the cold front caused the changing cloud pattern on Lookout Mountain and was responsible for attaching the meteorological connotation to the action: "The Battle Above the Clouds."

Sherman's March to the Sea: November–December 1864

It was "Yankee weather" in November and December during the twenty-five days required for General Sherman's march from Atlanta to the outskirts of Savannah. Only two periods of rainfall occurred to hinder the progress of this controversial action of war. Meteorological data for Georgia during this period were nonexistent, but contemporary diaries of participants confirm that the expedition was favored by the cool, dry conditions most conducive for marching.

The nearest meteorological station was located in the Union beachhead in southeastern South Carolina at the U.S. Medical Department's stations at Hilton Head and Beaufort, where daily readings were taken. Precipitation records in Confederate territory have been preserved for Camden to the northeast of Columbia, South Carolina, and at Aiken, on the Georgia border northeast of Augusta. Both locations were near the line of advance of Sherman's army. Unfortunately, only monthly totals and not daily entries have been preserved for these two stations.

The most perceptive, on-the-spot account of the march was composed daily by Colonel Henry Hitchcock of St. Louis, who was always aware of the importance of the weather factor for a military operation. He recorded only two rain periods during the twenty-five days: on November 19–20, the fourth and fifth days of the march; and on December 6–7, the twenty-first and twenty-second days out. The Medical Department records indicated a heavy rain during the first period: 1.50 inches at Hilton Head and 1.79 inches at Beaufort. The second rainy period mentioned by Hitchcock brought only 0.07 inch at Beaufort, and the Hilton Head record for all December is missing. Both November and December were very dry months at all South Carolina stations. Colonel Hitchcock summed up his opinion of the attendant weather conditions on December 10, the twenty-fifth and final day of the first leg of Sherman's march: "As these memoranda show, we have been most fortunate in weather — have had but two days of rain, one of cold (not severe) and one or two others only on which the weather was not everything we could wish."

After the capture of Savannah, Sherman's army took a long rest before heading north through the Carolinas on January 22. Heavy rains for several days had delayed the takeoff, but fine weather followed the rest of the month with several below-freezing nights. Hitchcock commented on February 2: "We have had so far a sufficiently 'agreeable march'; the weather fine, roads not bad on the whole as the distance shows." But conditions changed by the middle of February, with a period of heavy rains taking over. Many rivers of South Carolina went over their banks and the conditions were ever after known locally in the Camden area as the "Sherman Flood."

Appomattox Court House: The Storm Ends

The end to four years of fratricide came at Appomattox Court House in south-central Virginia on Palm Sunday, April 9, 1865, when Generals Grant and Lee agreed on terms of surrender. The actual weather on this day seems to be a matter of dispute.

No meteorological station was operative in the immediate vicinity, and no diary with a methodical entry of daily weather conditions has been located. Several diaries, apparently written on the spot, contain comments on local conditions. Most of the many accounts, written after a lapse of twenty to fifty years, describe the weather in a short line or two. Personal memoirs often represent only the hazy recollections of the participant or draw on what others have previously written, making them less trustworthy as sources of exact information as to what actually occurred.

Colonel Charles S. Wainright, who was present that day, wrote in his diary: "April 9. The day has been cloudy with rain at intervals. I have brought all my battery up . . . and have pitched my tents on the best bit of ground I could find. Bad enough it is with mud raised by this sopping rain . . . April 10. Monday. Another wet, nasty day."

Writing in 1954, historian Bruce Catton, who was not present, related in *A Stillness at Appomattox*: "It was Palm Sunday, with a blue cloudless sky, and the warm air had the smell of spring. The men came tramping up the fields by the railroad station with the early morning sun over their right shoulders . . ." And in midmorning: "The sun gleamed brightly off the metal and the flags."

For the events of the morning, Catton refers in his sources to *A Private's Reminiscences of the Civil War*, a memoir by the Reverend Theodore Gerrish published in 1882: "It was a beautiful spring morning; the air was soft and balmy; the sun shone from a cloudless sky, and as he climbed in the eastern horizon he saw two great armies in close proximity to each other."

Others describe the fog of the morning: "The early morning had been damp, slightly foggy, and presaged rain," wrote General George A. Forsyth in 1900; and "The fog of the morning was just rising from the open fields . . ." were the words of Henry Edwin Tremaine in *Last Hours of Sheridan's Cavalry* in 1904.

In what may be the definitive retelling of the Appomattox story, *An End to Valor*, Philip Van Doren Stern wrote in 1958: "The early morning of Palm Sunday, April 9, was foggy." He probably drew on the 1928 memoir of John A. Gibbon entitled *Recollections*: "The early morning had been damp, slightly foggy, and presaged rain." This is a paraphrase of Forsyth's words quoted above.

As for the morning conditions, the best evidence indicates that it was foggy. As to the noon hours, the only mention found came from the Forsyth memoir: "For while the day had developed into warm, bright, and beautifully sunny weather . . ." Most seem to agree that rain fell in the evening.

Our only indication as to what the large-scale weather was over the Virginia area came from the Naval Observatory at Washington. A high-pressure area was passing off the coast of Virginia on the ninth, the barometer having peaked at 30.40 inches (102.9 kPa) early in the morning. Skies were overcast all day. The thermometer stood at 53°F (12°C) at 3:00 P.M. Rain began at 9:00 P.M. and continued for twenty-seven hours, until midnight of the following day. According to the circulation pattern then existing with southwest winds at Washington, the weather at Appomattox should be quite similar to that at the nation's capital — hardly a bright, sunny, spring day!

CHAPTER THREE

Presidential Weather

The presidency of the United States is closely concerned with the state of the weather as well as with the state of the Union. Often the two are intertwined. On Election Day, current weather may be the decision maker in the personal plans of numerous citizens in visiting the polls or not. Next, on Inaugural Day, fine weather helps to enhance the pageantry of the ceremonies and starts a new administration with a mood of fair-weather optimism. During the four hard years in office the President and his advisors are called on to deal with natural disasters in the form of tornadoes, hurricanes, floods, and droughts. Since 1950 it has been the responsibility of the presidential office to designate disaster areas so that all arms of the federal government can be mobilized for relief and rehabilitation.

Close Elections and the Weather

Does the weather affect the outcome of a presidential election? To answer this question one must consider whether the voting took place in the horse-and-buggy days or in the period of hard roads and automobiles, whether the area under analysis was rural or urban, and whether the pollsters or politicians prejudged the election to be close or a landslide for one candidate. Also the efficiency of the local political organization in getting out the vote comprises a big factor. The answers to all these questions lie beyond the scope of this work. Our purpose here is to present what the actual weather conditions were in the crucial states where a change in a relatively few votes would have influenced the outcome of the national electoral vote.

One feature of the electoral panorama stands out in the span of almost two hundred years of presidential electioneering: the importance of New York State, with its large bloc of votes and its often even division in party affiliation. Since 1792 the Empire State has been on the winning side in all except six elections, and all but one of the misses occurred when one of its past, present, or future governors was the candidate.

New York presents a great contrast, not only between its urban concentration in the southeast sector and the predominantly rural upstate region, but also the climatic regime of the southeast versus upstate. The National Weather Service currently divides the state into no less than ten climatic zones. The northwestern and western parts lie in the Great Lakes region along the St. Lawrence Valley storm track, while New York City and Long Island fall under the influence of the Atlantic coastal storm track with its northeasters and tropical storms. Especially in November, weather conditions on a particular day may vary considerably between the western border of Chautauqua County, lying almost on the meridian of Pittsburgh, and the eastern tip of Long Island, on the meridian of southwest Rhode Island. And from the northern border with Canada to the southern tip of Staten Island, there is a span of 4°30' or about 315 miles. On an early November day the difference in mean temperature between Massena in the northwest and New York City in the southeast amounts to ten degrees Fahrenheit.

On any November morning a cold front might divide the state into different weather zones. Polar airstreams may cause blustery rain or snow showers in the Buffalo area, while the lower Hudson Valley is basking in Indian summer mildness. Moving at an average rate of 25 to 30 mi/h (40 to 48 km/h), the frontal system and attendant bad weather might not reach the New York Metropolitan area until the polls have closed.

On election day in the nineteenth century the afternoon

newspapers in New York City used to carry dispatches from around the state describing the expected turnout from the size of the early polling combined with an exact statement as to the existing weather conditions.

There is an adage in New York politics that a rainy day favors a Democratic candidate since the upstate Republicans would not turn out in full strength in inclement weather, while the urban Democrats would not be put to undue inconvenience in walking to their neighborhood polling places on hard pavement. The very close elections of the 1880s comprise the outstanding examples of New York State's holding the key to the presidency; a crucial shift of 10,517 in 1880, 575 in 1884, and 7189 in 1888 in the state's vote would have reversed the national electoral outcome. Local weather conditions during the three elections will be analyzed later.

Illinois, with a sizable bloc of votes, has always been a bellwether state, being on the winning side in all but five national elections since 1820. It also has a varied makeup, from urban concentrations in Cook County, along Lake Michigan, to the rural central and southern sections narrowing to "Little Egypt," at the confluence of the Ohio and Mississippi rivers. The state extends north and south about 5°30', or 370 miles, placing it in eight climate zones. The North partakes of the Upper Midwest weather pattern, along the main storm tracks from the central Great Plains to the St. Lawrence Valley. Cairo, in the southern extremity, lies in the latitude of Norfolk, Virginia, and enjoys many of the climatic features of the Border States, with an eight-degree temperature differential from the Chicago area. The storm track from the Gulf of Mexico over the Tennessee and Ohio valleys is the principal precipitation producer in November.

California, long a pivotal state, has been of growing importance. The spectacular population increase in recent decades made it the most populous in the Union, with an allotment of 47 electoral votes. The Golden State has been on the losing side only five times since 1852, and in two of those five the vote was divided between two candidates. It has the most varied climate of any state, with sandy deserts below sea level in close proximity to snow-clad peaks of 14,000 feet elevation and palm-bordered beaches within sight of northern conifers atop alpine mountains. In November there is a twenty-degree temperature differential between the extreme northeast and southeast.

In early November, weather systems moving inland from the North Pacific Ocean may give the northern counties steady rains, while the southern counties bask in sunshine, and the reverse may occur when cyclonic storms from the central Pacific Ocean affect only the lower third of the state. Often a frontal system causing rain in the North may take twenty-four hours to reach the South. The Democratic North might be undergoing a dousing, while the Conservative South enjoys sunshine on a November election day. In some seasons, however, the rainy season of winter has not yet been fully established during the first week of November, so a uniform weather pattern may prevail over the state and nullify the weather as a turnout factor at the polls.

Election Day Weather

Prior to 1824, there were no national election days when the people of all the states cast their ballots on the same day for the presidential candidates. Also prior to that time, no systematic collection of weather reports on a national scale was possible to enable one to make a surmise as to what the weather map of that day might look like.

Most elections since that time have been decided by a substantial margin, yet several were very close and the shift of a few popular votes in particular states would have changed the results. This was especially true in the elections during the 1840s, the 1880s, and in 1916, 1948, 1960, and 1976.

William Henry Harrison — 1840 —
Four Eastern States Critical

William Henry Harrison defeated incumbent Martin Van Buren by a narrow margin in the popular vote. A shift of only 8383 votes in four states (New York, Pennsylvania, Maine, and New Jersey) would have changed the outcome. On election day a high-pres-

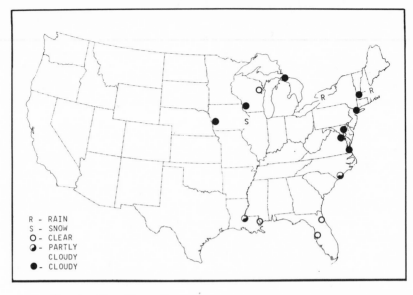

R - RAIN
S - SNOW
O - CLEAR
◖- PARTLY
   CLOUDY
●- CLOUDY

Election Day Map, 1824
Cloudy conditions prevailed over the East with an all-day rain at
Salem, Massachusetts, and rain near Buffalo. The Southeast and
Gulf States were mainly clear. Varied cloudiness prevailed over
the Great Lakes. Noontime temperatures: Salem, Massachusetts,
44°F (7°C); Baltimore, 54°F (12°C); St. Augustine, 70°F (21°C); Bat-
on Rouge, 78°F (26°C); and Council Bluffs, now Iowa, 56°F (13°C).

sure system dominated the weather from Ohio eastward, with
fair skies and seasonable temperatures favoring a large voter turnout in the crucial states.

James K. Polk — 1844 — New York State Decisive

The original "dark horse" candidate for his party's nomination,
James Polk also came from behind in the election to secure victory by capturing New York State by only 2555 votes over Henry
Clay, a tireless seeker of the presidency. A shift of only 0.097 of
one percent in the New York vote would have changed the national outcome. Election day weather was partly cloudy in the
western part of the Empire State and cloudy in the eastern part,
without any precipitation's being reported. Nothing in the

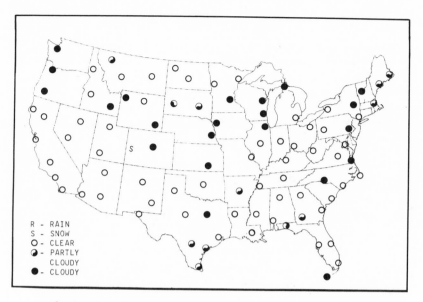

Election Day Map, 1916
A long cold front extending from Hudson Bay to a secondary
storm center over the Texas panhandle caused heavy rains in the
Midwest: Marquette, Mich., 0.86"; St. Paul, 0.82"; Des Moines,
0.82"; St. Joseph, Mo., 1.16"; Wichita, 2.06"; Altus, Okla., 0.42".
Rains were commencing in the Northwest: Tatoosh Island, Wash.,
0.34". The Atlantic seaboard basked under an extensive ridge of
high pressure from Maine to Florida.

weather scene held voters from going to the polls in the normal
manner, either in the Whig districts upstate or in the Democratic
strongholds in New York City.

James A. Garfield — 1880 —
New York Outcome Crucial

James Garfield won the election over Winfield S. Hancock with
only a minuscule one-tenth of a percentage point in the popular
vote, though he triumphed handily in the electoral college by 214
to 155. New York State proved crucial. Garfield won by only
10,517 votes and took its 36 electoral votes. A reversal of the New
York tally would have given the election to Hancock. Fine

weather, associated with an extensive high-pressure area, prevailed throughout the Empire State on election day, a factor that assisted Garfield in getting out the vote in the normally strong Republican districts in rural upstate.

Grover Cleveland — 1884 —
575 Votes in New York State Decisive

Bachelor Grover Cleveland was the first Democrat to be elected President since bachelor James Buchanan in 1856. At 300 pounds, he was also the largest man so far to win the presidency. The victory was another close affair: by 0.2 of a percent of the popular vote and by 37 votes in the electoral college. New York proved the pivotal state, where a shift of 575 votes would have elected James G. Blaine instead of Grover Cleveland. The weather at New York City "was good enough for election purposes," stated the *New York Post*. Upstate, "the weather was excellent" for a large Republican turnout at the polls. An inept remark by a clergyman in Blaine's camp about "Rum, Romanism, and Rebellion" had more to do with the defeat than the weather.

Benjamin Harrison — 1888 —
New York Pivotal Again

The seesaw battle for the presidency that marked the last quarter of the nineteenth century found rivals with worthy credentials in the election of 1888. Incumbent Grover Cleveland was opposed by Benjamin Harrison, the grandson of a President and great-grandson of a signer of the Declaration of Independence. Though Cleveland won a greater popular vote, Harrison carried the 36 electoral votes of his opponent's home state, New York, and won the election. A shift of only 7189 votes would have returned Cleveland to the presidency. On election day, the Empire State occupied a post-frontal situation as a storm system passed eastward into New England. A cloudy morning gave way to clearing in the afternoon, presenting little weather deterrent to keep voters from going to the polls.

Woodrow Wilson — 1916 —
Late California Returns Reversed Outcome

Weather probably played a greater influence in Woodrow Wilson's second election than in any other presidential contest. A shift of 1983 votes from Wilson to Charles Evans Hughes in California would have changed the national result. On election day morning, California was emerging from its first general storm of the winter season. From 15 to 20 inches of snow lay on the higher mountains, and the temperature at Donner Summit, where the railroad crosses the High Sierra, dropped to 10°F (−12°C). The snow area covered all the mountainous counties in the northeast corner of the state, which were generally Democratic in allegiance, and these were the last to report their tallies. The outcome became a cliffhanger until late the following day because the snow had downed some telegraph lines in remote districts. Wilson emerged with a majority of about 2000 votes, giving him the state and the election. Without the storm, Wilson might have garnered a greater margin. But if the storm had been more severe or continued a few hours longer, fewer votes might have been cast in the mountainous districts and his narrow margin dissipated.

Harry S Truman — 1948 —
Late Returns Spell Victory

Though the early morning edition of the *Chicago Tribune* announced the victory of Thomas Dewey, late returns gave Harry Truman enough electoral votes to win by 303 to 189. A shift of 29,294 votes (0.28 of one percent) in three states (Illinois, California, and Ohio) would have made Dewey the next President. A switch of 12,487 votes (0.2 of one percent) in California and Ohio would have denied both candidates an electoral majority and thrown the election into the House of Representatives.

A storm system in the Lower Mississippi Valley spread rain over all of Illinois on election day; amounts were heavy in the South with some stations reporting over 2 inches. No rain fell in Ohio until after 7:00 P.M., when it commenced in the Cincinnati area. Skies were generally cloudy. The temperature at Columbus ranged between 51°F and 68°F (11°C and 20°C).

Northern California received rain in the morning and again in the evening after 7:00 P.M. from a Pacific storm that was just entering the Washington coast. San Francisco measured 0.31 inch, but Los Angeles had none. The temperature at Sacramento ranged from 55°F to 71°F (13°C to 22°C). The weather was a factor in this election chiefly in Illinois, only secondarily in northern California, and not at all in Ohio.

John F. Kennedy — 1960 — Narrow Victory Hinged on Five States

Though Kennedy had a seemingly adequate margin of 84 votes in the electoral college over Richard Nixon, a shift of only 11,424 popular votes in five states (Illinois, Missouri, New Mexico, Nevada, and Hawaii) would have elected Nixon. A cold front advancing from the Great Plains caused a steady rain on election day in Illinois and Missouri. This would have deterred Republican voters in the rural areas in getting to the polls, but not cut down substantially in the Democratic turnout in Chicago and Cook County, whose votes determined the Illinois outcome. The weather in the other three states was favorable for voters.

Jimmy Carter — 1976 — Closest Since 1916; 3800 Votes Decisive

Jimmy Carter had a margin of only 54 electoral votes over incumbent Gerald Ford. A reversal of only 3800 votes in each of Ohio and Hawaii would have changed the outcome and given Ford the victory. It was the closest election since 1916. A total of 174 electoral votes were decided in states where a 2 percent or lower difference in the popular vote was tallied.

Ohio enjoyed clear skies on election day. In Hawaii, little or no rain fell during the day except on the windward slopes of the mountains. Honolulu city and airport were rainless. Thus, the weather could not be included in the reasons causing Ford's close defeat.

Inauguration Day Weather

No sentence in our language has more portentous content than the thirty-five words composing the oath of office of the President of the United States. The simple statement, set down in Article II, Section 1, of the Constitution, has been spoken without change by the thirty-nine persons who have held the office of chief executive. Though the scene since 1817 has usually been the east portico of the Capitol, on special occasions it has ranged from Wall Street in New York City to a remote Vermont farmhouse, and from *Air Force One* in Dallas, Texas, to the White House itself. The chief varying factor in all inaugurations has been the weather overhead. There have been beautiful interludes of "false spring" when sunny skies have driven the temperature up to the high 50s and even low 60s, but on other occasions driving snowstorms, bitter wind, or pelting rainstorms have forced the ceremonies indoors or curtailed the parade.

The rituals attending the swearing in of the President of the United States comprise the nation's most impressive display of ceremonial activities, combining both solemnity and gaiety. From the morning hour when the outgoing President leaves the White House for his last ride down Pennsylvania Avenue until the incoming President returns from the Inaugural Ball in the small hours of the next morning, Washington is the center of the heart and thoughts of the entire nation, and now through the electronic magic of satellite television most of the countries of the world share in the spectacle of Washington's day of pomp and circumstance.

The choice of the date of March 4 for the commencement of a new administration was a decision of the last Congress under the Articles of Confederation. The institution of the new government in early 1789 moved slowly. The elected congressmen gathered "without haste" for the scheduled meeting on the first Wednesday in March. It was not until early April that a quorum was present to count the electoral ballots and declare the election of George Washington.

The first President of the United States took the oath of office at Federal Hall in New York City on April 30, but four years later he reverted to the originally planned March 4 date for his second inauguration at Philadelphia. On this date, or the day following, it remained fixed for 144 years despite some very inclement encounters with the atmospheric elements at this season when the weather lion traditionally roars.

The original date of March 4 was established to permit the newly elected President to make arrangements for leaving home, travel by primitive transportation to the capital, and assemble a staff. The journey from southwest Georgia or from the northern part of the District of Maine might take as long as three weeks of overland travel. The extreme localities of the original thirteen states happen to be almost equidistant, 750 miles, from the proposed District of Columbia.

With the speedup in the tempo of national life brought by radio and air travel, however, Congress in 1932 decided on a "New Deal" in presidential and congressional arrangements. An amendment was proposed to the Constitution to advance the date of the inauguration of a new administration forward to January 20 to speed up the transfer of governmental reins to the incoming President with a popular mandate. The "lame duck" aspect of the situation was primary in the move, though the weather factor was thoroughly investigated at the time. Data showed that January 20 had a better record, stormwise, than March 4 in the Weather Bureau record books. But the first inauguration on the new date, Roosevelt's second, in 1937, brought the heaviest rainstorm ever experienced by an incoming President, 1.77 inches, dampening the ardor of even the staunchest New Dealer.

Said Senator George W. Norris of Nebraska, author of the Twentieth Amendment, "They're trying to blame this one on me. You can't charge this up to me until after March 4, when you see what kind of a day that is." It turned out to be a beautiful spring day — sunny with a high temperature of 67°F (19°C), unusually warm for an early March date.

Weatherwise, the change from March 4 was unfortunate. A survey of Washington weather on that date from 1941 to 1981 indicates only one day with unfavorable conditions. In 1965, the

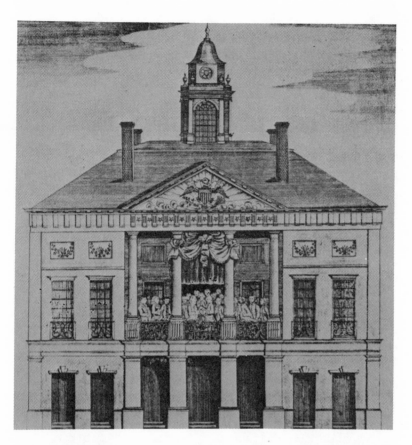

The first inauguration. President Washington on a balcony of Federal Hall in New York City. Clear 59°F, NW wind. Courtesy of the Library of Congress.

year of Johnson's inauguration, March 4 was the middle day of a three-day rainstorm: 0.42 inch fell of a storm total of 1.98 inches. Kennedy's inaugural shared the fate of Roosevelt's second; instead of snow and cold on January 20, Kennedy would have enjoyed 68°F (20°C) temperatures and no rain on March 4. If the first seven inaugurals, from 1789 to 1817, enjoyed favorable weather conditions, so too would the last eleven, from 1941 to 1981, with the exception of 1965, have enjoyed similar fair weather if held on the originally prescribed date of March 4.

George Washington — 1789 — Bright Spring Day

President-elect George Washington took the oath of office at about one o'clock on Thursday, April 30, 1789, on the balcony of Federal Hall, overlooking Broad and Wall streets in New York City. The citizens of New York had contributed $32,000 to transform the old City Hall into Federal Hall, whose open balcony above the street made a dramatic stage. After ending the oath with an added "I swear, So Help Me God," the President returned to the Senate Chamber, where he delivered his inaugural address. It was a "bright spring day," in the words of inaugural historian Louise Durbin.

According to the meteorological register of Henry Laight, the skies at noon were clear, the temperature stood at 59°F (15°C), and the wind blew from the northwest — conditions indicative of a fair-weather, high-pressure system positioned to the west of New York City. Henry Laight, who was employed by the New York Insurance Company, began weather observing in 1788 and continued until 1836, when he broke his thermometer.

George Washington — 1793 — Hazy Sunshine, Mild

In sharp contrast to the exuberance of the first, Washington's second inauguration, on Monday, March 4, 1793, was a much simpler affair. The federal government had moved to Philadelphia on a temporary basis while the new federal city on the Potomac River was being prepared. At a few minutes before noon, President Washington arrived at Congress Hall on Independence Square, where the ceremonies were held indoors.

The meteorological record of David Rittenhouse, the astronomer and instrument-maker, tells us that the weather in the early afternoon was characterized by hazy sunshine. The temperature was a balmy 61°F (16°C), with a south wind blowing.

John Adams — 1797 — Rain Early, then Fair

John Adams became the second President of the United States on Saturday, March 4, 1797, in the chamber of the House of Representatives of Federal Hall in Philadelphia. Though David Rittenhouse died in 1796, members of his family maintained the weather observations for several years. The entry for this date indicated: "Cloudy, some rain A.M., fair P.M." The thermometer read 53°F (12°C), with the wind coming out of the southwest at the early afternoon observation.

Thomas Jefferson — 1801 — Mild and Beautiful

It is ironic that the inauguration of the only President who might be classified as a meteorologist marked the single instance that no indication of local weather conditions was available. Traditionally, Tuesday, March 4, 1801, was supposed to be "mild and beautiful," according to an 1898 congressional study of the advisability of changing the inaugural date to later in the spring. "The occasion was favored with a beautiful day, as the sun shone brilliantly and the weather was mild," in the words of Thomas H. McKee, a chronicler of elections and inaugurations writing in the 1890s, but he gave no source for his statement.

We do know what the weather was on this date at moderate distances from Washington. At Orange, Virginia, some 70 miles to the southwest, the meteorological records of the James Madison family indicated fair skies, a temperature of 58°F (14°C), and an east wind prevailing. To the northeast, at Philadelphia, about 120 miles distant, the Rittenhouse record also reported fair skies, a temperature of 47°F (8°C), and a southwest wind. An analysis of these reports would suggest that no weather front existed between Pennsylvania and Virginia and that Washington most likely shared the favorable conditions and a temperature in the low 50s.

Favorable conditions are also indicated by the fact that Jefferson walked from his lodgings on New Jersey Avenue to the Capitol, and that in the evening celebrants set off "a pretty general illumination."

The editor of the *Times and District of Columbia Advertiser* of Alexandria, Virginia, commented on the day: "It affords us pleasure to add, that notwithstanding the vast concourse of people at the city, not a single accident occurred to damp the joy so universally diffused upon the happy occasion."

Thomas Jefferson — 1805 — Fair and Temperate

Soon after moving into the White House, Jefferson installed his thermometer, which he had purchased in Philadelphia on July 2, 1776, and maintained a daily record of the weather elements. He certainly deserves the designation of "First Chief of the Weather Bureau" or "First Director of the National Weather Service," as the position is currently called.

On the morning of Monday, March 4, 1805, the White House thermometer read 32°F (0°C) and the sky condition was fair. This was the only entry for that day since Jefferson was busy elsewhere in the city later on. The Philadelphia record also indicated fair conditions at noon, with a temperature of 52°F (11°C). The next day, too, was fair, indicating the presence of anticyclonic conditions and a spell of good weather over the Middle Atlantic area.

James Madison — 1809 — Cloudy After Rain, Seasonable

"The scene was truly pleasing and magnificent. The day was cloudy, which with the rain which fell the evening before, rendered it disagreeable," wrote an anonymous correspondent from Washington to the *Maryland Gazette* at Annapolis.

Thomas Jefferson recorded a reading of 40°F on the afternoon of March 3, but then packed his thermometer for the return trip to Monticello. His entry on the morning of March 4 was simply: "car," meaning cloudy after rain in his code. No doubt, the streets were very muddy for the ten thousand people who gathered around the Capitol to see James Madison inaugurated.

James Madison — 1813 — Sunshine and Fresh

The country was at war with Great Britain on Thursday, March 4, 1813, when James Madison took his second oath of office. "The day was fine; the sun shone brilliantly as if to welcome it," recorded the Washington *National Intelligencer*. Expanding on this,

Thomas McKee wrote: "The seventh quadrennial day of the United States was ushered in with all the freshness and beauty of an early spring morning, all nature seeming to welcome the great event."

With the absence of the Jefferson thermometer, no other instrumental record has been found either in newspapers or in diaries.

James Monroe — 1817 — Clear and Mild

The Era of Good Feeling was ushered in, appropriately, with good weather. "The day was auspicious beyond example. The air was calm — the sun warming and invigorating. About eight thousand people assembled for the occasion," reported the correspondent of the *Genius of Liberty*, a newspaper published at nearby Leesburg, Virginia. "A delightful spring day," was the description of the *Georgetown Messenger*. Though no meteorological report for Washington was available, at Baltimore the record of a Captain Brantz described "the clear and mild" conditions there with a thermometer reading of 48°F (9°C) at his early afternoon observation.

James Monroe — 1821 — Early Snow, Cloudy, Cold

For the first time since the founding of the Republic, the inaugural ceremonies experienced bad weather. It may have been because March 4 fell on a Sunday, and President Monroe established the precedent of pushing the inaugural over to Monday. "The day proved very unfavorable for the attendance of spectators, there having fallen during the night a good deal of snow and rain," reported the *National Intelligencer*. Captain Brantz measured an inch of snow as falling during the night at Baltimore, and then the weather cleared later in the morning of the fifth. His morning temperature reading was 33°F (1°C), indicating that some of the snow must have melted and made roads muddy.

A series of local weather observations started in 1821 at a location somewhere on the side of Capitol Hill. They were carried on by members of the Columbian Institute, a local scientific and cultural body serving as a predecessor to the Smithsonian Institution. Professor Josiah Meigs, a former educator at Yale and the University of Georgia, was a leader of the group and a meteorological enthusiast, and it is thought that he initiated the series of observations. The record shows that the snow began the previous evening at 7:00 P.M. and was still filtering down at sunrise of inauguration day. Winds were out of the northeast, indicating a coastal storm. At 2:00 P.M. on inauguration day the thermometer read 34°F (1°C) and the sky was cloudy. A barometer reading of 29.74 inches (100.7 kPa) was indicated, with no hour specified.

John Quincy Adams — 1825 — Early Rain, Cloudy, Cool

The incoming President, himself, supplied the meteorological details for his inauguration. As secretary of state, John Quincy Adams maintained a weather diary that usually contained the sky condition, temperature, and amount of precipitation.

The Reverend Robert Little, the current observer for the Columbian Institute on Capitol Hill, had three observations on March 4, 1825. His 2:00 P.M. readings reported the sky covered with nimbus clouds and the temperature at 46°F (8°C), having risen only 4°F (2°C) since morning. The barometer read 30.10 inches (101.9 kPa), and the wind was out of the northeast. A total of 0.79 inch of rain fell during the day.

Andrew Jackson — 1829 — Clear and Mild

"The day was serene and mild in every way favorable to those who had come from a distance to witness the ceremony of the Inauguration," reported the *National Intelligencer* about the weather on Wednesday, March 4, 1829. The *United States Telegraph* confirmed this, saying: "The day itself was remarkable and pleasant, and all nature seemed to rejoice with the people."

William Elliott, now the observer on Capitol Hill, recorded a maximum temperature of 61°F (16°C), with skies clear and a

Jackson on his way to Washington in 1829, as drawn by Howard Pyle for the 1881 Harper's Weekly. *Courtesy of the Library of Congress.*

light breeze blowing from the southwest. Jackson was favored by the most pleasant inaugural day so far.

Andrew Jackson — 1833 — Clear and Cold

Andrew Jackson's second inauguration proved a much less tumultuous affair when held in the House of Representatives on Monday, March 4, 1833. Snow covered the ground that morning on a bright, winter day. All plans for a public demonstration were canceled as a result of the low temperatures and the condition of the 65-year-old President's health. The *Washington Globe* announced in its morning edition that "in consequence of the inclemency of the weather, the ceremonies of the Inauguration will take place indoors."

Mrs. William Thornton noted in her diary: "Fine day for the inauguration — cold."

The Capitol Hill weather observer entered a thermometer reading of 26°F (− 3°C) at sunrise, and it rose to only 30°F (− 1°C) by midafternoon. The sky continued clear with a brisk breeze from the northwest sweeping the city — true anticyclonic conditions with a high-pressure system located to the northwest.

Martin Van Buren — 1837 — Hazy Sunshine with Clouds, Cold

Martin Van Buren of New York became the first President to be born under the flag of the United States. Though his inauguration day was serene, his administration proved stormy when the Panic of 1837 hit soon after his taking office.

Saturday, March 4, 1837, was described by the *Globe*: "A lovely day of brightest sunshine gladdened every heart — a soft spring snow, which had fallen two days before, lost, in the warm day, its touch of water, and, in virgin purity, reflected from the surrounding hills the cheering light and benignity of the Heavens . . ." The ceremonies were held on the east portico of the Capitol.

The temperature on Capitol Hill dipped to 9°F (− 13°C) at sunrise and rose to a maximum of 28°F (− 2°C) in the afternoon. Sky conditions were described as "cloudy, hazy, little breeze." The high reading of the barometer at 30.48 inches (106.6 kPa) and the northeast breeze indicated that the Washington area lay under the influence of a strong anticyclone, insuring good weather.

William Henry Harrison — 1841 — Some Clouds, Seasonable

At sunrise on March 4, 1841, a 26-gun salute — one gun for each state — was fired on the Mall in front of the Capitol to herald the day when William Henry Harrison, "Old Tippecanoe" of the War of 1812 fame, would assume the presidency. The 68-year-old national hero, the winner of the riotous "Log Cabin and Hard Cider" campaign, was the oldest man to be elevated to the office so far and would serve the shortest term — one month. He rode without an overcoat in the parade, and the chill he received led to a cold and eventually to his death of pneumonia.

"The morning broke somewhat cloudily, and the horizon seemed to betoken snow or rain," observed the *National Intelligencer*, as if the weather seemed a premonition of things to come.

The new Department of Charts and Instruments of the U.S. Navy, located on the side of Capitol Hill, now served as the official weather station. The observer noted an early morning temperature of 34°F (1°C) and a midafternoon reading of 51°F (11°C). The morning sky went from clear to cloudy, but by afternoon only high-altitude cirrus clouds were seen. Northeast to east winds of moderate strength prevailed during the daylight hours. No precipitation fell on the fourth, though a snowstorm started at 2:00 P.M. on the fifth.

John Tyler — 1841 — Clear and Mild

Vice President John Tyler was at his home in Williamsburg, Virginia, when Fletcher, the son of Daniel Webster, and another member of the State Department arrived at 7:00 A.M. with the sad news of the passing of the senior partner of the election team of "Tippecanoe and Tyler, too." William Henry Harrison died just a month to the day after assuming the high office. John Tyler departed immediately for Washington. The oath of office was administered at noon on April 6, 1841, indoors at Brown's Indian Queen Hotel. The weather that day, according to the records of the Naval Depot of Charts on Capitol Hill, was generally clear after a rainy night. A cold front passed through the area sometime before 9:00 A.M., turning the wind from light south by west to fresh northwest. The temperature rose to 55°F (13°C), and with a rising barometer moderate west winds prevailed the rest of the day.

James K. Polk — 1845 — Steady Rain

James K. Polk's inauguration was the darkest of record so far. The ceremony was dubbed "an assemblage of umbrellas," and it was

Polk's inaugural in 1845 as viewed by a correspondent of the London Illustrated News. *Courtesy of the Library of Congress.*

said that the noise of the rain beating on these prevented the inaugural address from being heard. Despite the downpour of rain, inventor Samuel F. B. Morse was able to send a message by telegraph, perfected only a year before, from the Capitol platform to Baltimore, setting a record for the speedy reporting of an inaugural.

The *National Intelligencer* described the scene:

> But the hopes of the morrow, which had kept many an eye waking through the preceding night, were all sadly dashed by the unrelenting, undescriminating sky, which after a transient smile, began to lower, and frown, and finally to pour down rain outright . . . and, accordingly, such a display of umbrellas as darkened the city by their shade was never probably witnessed by "the oldest inhabitant."

The U.S. Navy's weather station was now located at the site of the Naval Observatory on E Street, overlooking what is now

Potomac Park, close to the present Lincoln Memorial and the Kennedy Center. The only observation taken on inaugural day was at 9:00 P.M., when the temperature read 50°F (10°C), a light wind blew, and it was raining. The all-day rain totaled 0.40 inch.

Zachary Taylor — 1849 — Cloudy, Cool, Snow Flurry

General Zachary Taylor had never voted in a presidential election until he did so as a candidate in November 1848. He was 66 years old and had been through many military campaigns, earning the title of "Old Rough and Ready."

March 4 fell on a Sunday, so the inaugural ceremony was put over to Monday. "The weather was, upon the whole, though the sky was clouded, as pleasant as could have been looked for in this particular season," observed the *National Intelligencer*. "Fine weather and an unlimited noise and festivities characterized the inauguration," commented historian Thomas McKee.

The weather at noon at the Naval Observatory showed a temperature of 42°F (6°C), an overcast sky, and a light wind from the east. It snowed briefly around 3:00 P.M.

Millard Fillmore — 1850 — HOT, HOT, HOT

Among other unenviable distinctions, Millard Fillmore endured the hottest oath-taking day of any President of the United States. Upon the death of Zachary Taylor, who suffered a heat exhaustion at the dedication of the Washington Monument, Vice President Fillmore was sworn into office on July 10, 1850, in the hall of the House of Representatives.

The thermometer at the Naval Observatory rose from a warm start of 74°F (23°C) to a maximum of at least 92°F (33°C) during the afternoon. There was no registering thermometer to catch the actual peak; observations were taken at three hourly intervals. Skies were overcast at the noon hour, but breaks in the clouds with eight-tenths coverage developed by midafternoon. Winds came from the east in the morning, but shifted to south by afternoon, movement being light at all times. A moderate shower of rain in the amount of 0.33 inch had fallen by 4:00 A.M.

Franklin Pierce — 1853 — Raw, Windy, Light Snow

Franklin Pierce brought some of his New Hampshire weather to Washington for his inauguration. The *National Intelligencer* commented: "The weather was not pleasant; a raw northeasterly wind, wafting a pretty continuous though melting snow, made its effects felt even on young blood when not kept in motion; still it was not forbidding enough to prevent any but invalids from giving themselves to the scene in the open air." The *New York Times* correspondent added: "Snow continued falling slightly during the day, melting as it fell, and not particularly interfering with the inaugural ceremonies."

A noon report was not made that day at the Naval Observatory because of the many visitors who had to be shown around the building. The temperature at 10:50 A.M. and also at 2:50 P.M. was 35°F (2°C); with cloudy conditions existing, the noon reading must have been very close to that figure. Snow was falling with a light wind from the north and northeast. Robert F. Sems, the officer of the day at the observatory, remarked on the record sheet: "Genl. Franklin Pierce inaugurated as President of the United States of America. A proud day for the Demmys' this, But I could not be present. I suppose I was very much missed in the crowd."

Abigail Fillmore, the wife of the outgoing President, caught a cold while attending the ceremonies and died a month later.

James Buchanan — 1857 — Mild and Balmy

"No public pageant was ever favored with a more propitious sky than which yesterday signalized the Inauguration of JAMES BUCHANAN, fifteenth President of the United States," commented the *National Intelligencer*. Everything was beautiful that day, which was to mark the last time that a president of the Democratic Party was to be elevated to the White House for twenty-eight long years. Though winning only 45.6 percent of the popular count, Buchanan had split the votes of his Republican and Whig opponents, and won the electoral college. It was the calm before the storm.

President Buchanan and President-elect Lincoln share a carriage on the way to the unfinished Capitol in 1861. Courtesy of the Library of Congress.

"Two or three preceding days had been boistrous and cold, but both of these qualities abated during the night of the 3d, and the 4th came in mild and balmy, with an atmosphere calm and hazy, very much like that which distinguishes Indian summer," continued the *Intelligencer*.

The record of the Naval Observatory backed up these lyrical descriptions. At noon the sky was covered with three-tenths high cirrus clouds. A light wind blew from the west. The thermometer read 49°F (9°C) and went higher in the afternoon.

Abraham Lincoln — 1861 — Windy and Dusty

> The important 4th of March, 1861, dawned rather inauspiciously with leaden skies, and tornadoes of dust, which was leveled somewhat later by a slight fall of rain. As the morning moved on however, the skies brightened and the wind lulled, auguries noted with some complacence by those who pin their faith upon such omens.
> — *Evening Star*

No inauguration ceremony has been held under tighter security. Seven Southern states had already seceded from the Union and formed the Confederate States of America, with President Jefferson Davis installed in Alabama. The Confederate flag flew just across the Potomac River at Alexandria, within sight of the Capitol. Federal troops with fixed bayonets lined the route of the presidential procession, soldiers with rifles ready were posted atop buildings, and side streets were blocked with artillery ready to fire. Fortunately, nothing happened to interfere with the ceremonies except the dust.

"The dust was stifling as the procession neared the White House. The President, marshal and subalterns, the swells and the populace, were alike enveloped in it. One could have written a certificate of good behavior on the back of President Lincoln's coat as he entered the House," observed the *New York World*.

A cold front passed through the District about 12:25 A.M., attended by a wind shift to west and gales throughout the morning. At 10:05 A.M., the wind was clocked at 25 mi/h (40 km/h) and post-frontal clouds covered nine-tenths of the sky, according to Naval Observatory records.

Abraham Lincoln — 1865 — Clearing Overhead, Mud Underneath

Instead of dust, mud greeted the throngs who attended President Lincoln's second inauguration. The *Evening Star* put it: "This year the streets were covered with a thick coating of mud, carrying out the saying that Washington alternates, from dust to mud, or visa-versa."

As happened four years before, a windshift attending the passage of a cold front swept through Washington about sunrise. A brief, but moderately heavy shower of rain measuring 0.30 inch dropped enough moisture to cause muddy streets. Spectators were confined to the sidewalks, and the Corps of Engineers considered laying pontoon bridges along Pennsylvania Avenue for the afternoon parade.

Until noon, cumulus and cirrus clouds covered the sky to the extent of nine-tenths, but drier air soon arrived by the time the President delivered his address. At 3:00 P.M., only two-tenths of the sky was obscured, according to the Naval Observatory records. The noontime temperature was 46°F (8°C).

Walt Whitman, then a resident of Washington working in military hospitals, was on hand that day to pen his impressions of the weather and the scene in his inimitable style:

> On Saturday (March 4) a forenoon like whirling demons, dark, with slanting rain, full of rage, and then the afternoon, so calm, so bathed with flooding splendor from heaven's most excellent sun, with atmosphere of sweetness; so clear it showed the stars, long, long before they were due.
>
> As the President came out on the Capitol portico, a curious little white cloud, the only one in that part of the sky, like a hovering bird, right over him.

Andrew Johnson — 1865 — Dreary Rain

President Lincoln was shot at Ford's Theater on Tenth Street about 10:15 P.M. on Good Friday, April 14, 1865, while attending a performance of Laura Keene in *Our American Cousin*. The President died at 7:22 A.M. the following day.

The evening of April 14 was mild, with the sky only three-tenths covered and a gentle wind blowing out of the south. The thermometer at the Naval Observatory read 54°F (12°C).

Vice President Andrew Johnson took the oath of office at Kirkwood House on a rainy, drizzly day. The temperature at noon stood at 57°F (14°C). The wind was from the south, increasing from gentle in the morning to fresh in the afternoon. The rainfall, beginning at 7:00 A.M., amounted to 0.35 inch — in all, probably the dreariest day in the long history of our capital city and nation. The record sheet at the Naval Observatory contained the following note: "The building was draped in mourning for the death of the President of the United States."

Ulysses S. Grant — 1869 — Rain, then Sunny, Cool

General Ulysses S. Grant became the first of the line of Civil War heroes to be elevated to the presidency. Despite his appearance of maturity, Grant was only 46 years of age at his inauguration, younger than any of the seventeen men who had preceded him in the presidential chair. Though without previous political experience, Grant was to be the only President from Jackson to Wilson — a period of 76 years — to be elected twice and to serve two consecutive full terms in office.

The weather was described in the *Evening Star*: "The day opened rather gloomily, with lowering skies and occasional showers which settled into a steady drizzle. At ten o'clock Jupiter Pluvius began to relent and the opening skies gave promise of a fair day." The *National Intelligencer* added: "The afternoon was enlivened by a glorious sunlight, and people were in a high state of enjoyment."

The Naval Observatory noted rain falling lightly from 4:50 A.M. to 11:50 A.M., amounting to only 0.11 inch. The thermometer at noon reached 40°F (4°C), with the wind having shifted from southeast and now blowing lightly from the northwest. A weak cold front passed through Washington during midmorning. The sky at noon had nine-tenths cloud coverage, indicating that some sun was coming through the retreating cloud deck. By 3:00 P.M. only five-tenths cloud coverage remained.

Ulysses S. Grant — 1873 — Bitter Cold and Windy

In carrying out a resolution of Congress in 1870, President Grant had been instrumental in establishing a federal storm-warning service. This was placed under the direction of the United States Army Signal Service, where it continued for twenty years until its functions were assumed by the United States Weather Bureau under the Department of Agriculture. The Washington weather-men did not treat their creator with consideration upon his sec-ond inaugural. The *New York World* reported:

> While it has been clear, the weather has been anything but favor-able for the inaugural ceremonies. Last night was one of the coldest of the season, and this morning the window panes were orna-mented with the frescoes of Jack Frost. "Old Probabilities" was at fault for once in predicting a moderation of the high winds which prevailed yesterday, for it has blown a perfect gale to-day from the southwest. It is bitter cold, too; the fierce wind has pinched ears, twisted noses, and grasped thumbs as though determined to drive everybody within doors. Nor were people in haste to venture out. They lingered over breakfast and swallowed their steaming coffee with more than usual satisfaction, and convulsively shivered as the wind shrieked without the shutters, bearing everything movable along with it in its mad career. The few early birds who faced the wintry blasts went scurrying along with coat collars turned, hats pulled down, and icicles clinging to moustache and whiskers. Clouds of dust swept down the streets like the simoon in the des-ert, and against the blue sky the gay flags fluttered angrily, as if in protest at being exposed at such a time.

Clear skies, a strong northwest wind, and bitter cold pre-vailed. The official thermometer rose from a minimum of 4°F (−16°C) in the morning, to a noon 16°F (−9°C), and up to an after-noon maximum of 20°F (−7°C). The barometer climbed rapidly all day as the advective blasts of an approaching Canadian high-pressure area swept the District. The minimum of 4°F was not only the coldest inauguration day temperature, but still stands as a month-of-March record for Washington. The wind reached a peak of 23 mi/h (38 km/h) at the new U.S. Army Signal Service Observatory on G Street, N.W.

Rutherford B. Hayes — 1877 — Cloudy, Cool, Some Snowflakes

Things happened rapidly for Rutherford B. Hayes during the early days of March 1877. He left Columbus, Ohio, on the first not knowing whether he was actually President-elect; on March 2 the House of Representatives certified that he had won the disputed election by the margin of one vote; during the evening of March 3, he took the oath of office twenty-four hours before the scheduled time so that there would not be a hiatus in the presidency following Sunday noon; and on Monday, the fifth, he was again sworn in by Chief Justice White.

The weather on March 5 was described by the correspondent of the *New York Tribune*:

> The weather, though colder than residents of Washington have been accustomed to the last month, was very pleasant for this season of the year. The sun now and then struggled out through the clouds and tempered the breeze, which, though not strong, was raw and occasionally brought with it a stray flake of snow.
>
> ⋆ ⋆ ⋆
>
> The day dawned through clouds and cold . . . soon the sun came out and lit up the majestic white edifice [at oath-taking time] . . . The evening was beautifully pleasant.

The Signal Service office reported the weather as cloudy with brief periods of light snow and thermometer at 35°F (2°C).

James A. Garfield — 1881 — Partly Cloudy, Chilly

The weather at dawn on James A. Garfield's inauguration day did not augur well for the presidency that was to end tragically in a short six months. According to the *Washington Post*: "A heavy storm of wind, snow, and rain which commenced Thursday evening [the third] continued without intermission during the night. A more dismal appearing city than Washington at daybreak yesterday morning could not be imagined. A steady northwest wind, however, drove away the clouds, and by 11 o'clock the sun was shining brightly and the concrete pavement of the Avenue was in excellent condition."

After a night of rain mixed with snowflakes, a cold front passed through the city in the early daylight hours and the skies cleared in midmorning, giving an average cloudiness for the day of only two tenths. The temperature dropped gradually from a morning 35°F (2°C), to a noon 33°F (1°C), and to an evening 29°F (−2°C). The wind blew steadily from the west all day.

Chester A. Arthur — 1881 — Fair

While vacationing at his home in New York City, Vice President Chester A. Arthur was notified of President Garfield's death on the evening of September 19, 1881, a day described by the New York City weather office as having a "fair weather sunset." The temperature at 9:00 P.M. stood at 75°F (24°C). Arthur took the oath of office in New York City soon after midnight. On the twenty-second he traveled to Washington and again repeated the oath.

Grover Cleveland — 1885 — Partly Cloudy, Mild

Grover Cleveland enjoyed a partly cloudy and calm day for his first inauguration. The sky was covered with broken clouds during the morning and early afternoon, with clearing in the evening, according to the weather station records. At noon the temperature read 54°F (12°C). The wind was light from the north during the afternoon.

The *New York Tribune* correspondent described the scene on inauguration day:

> At dawn the sky was overcast and persons who failed to notice that the smoke rose in straight columns from chimneys feared the spell was broken and that two days of genial weather would be succeeded by at least a drizzle . . . The sun broke through the thin and drifting clouds and shone with mild energy on a swarming city . . . The rays of the sun beating through a hazy atmosphere with full force upon the platform made it uncomfortably warm, except when the breeze from the river would freshen and remind one that it was still March.

Benjamin Harrison — 1889 — Cold Rain All Day

Inauguration day for Benjamin Harrison proved to be the wettest since that of James Polk, forty-four years before. The *New York Tribune* dispatch described the conditions:

> Today has been distinguished by the wettest rain that ever fell alike on just and unjust. Moreover, the all pervading quality of a

A rainy parade from the Capitol after Benjamin Harrison's inauguration in 1889. Smithsonian Institution Photo Nos. 56995.

March wind has pierced to the marrow, however snugly the bones were overlaid with adipose tissue, and the man who wore rubber boots and peered forth from between the lapels of a Mackintosh has been the envy and despair of all beholders.

Rain began at 5:50 A.M. and fell heavily; a total of 0.52 inch was measured up to 8:00 A.M. A lighter fall continued throughout the day and well into the night, for a storm total of 0.86 inch. The temperature ranged from a low of 34°F (1°C) to an afternoon high of 44°F (7°C). The reading at 2:00 P.M. close to the time of the swearing-in ceremony was 41°F (5°C).

Grover Cleveland — 1893 — Rain and Snow and Cold

The second inauguration in a row with atrocious conditions added to March fourth's unsavory reputation. In the 1892 election, Cleveland had turned the tables on incumbent Benjamin Harrison, winning both the popular vote and the electoral college with comfortable margins. And the weather on inauguration day also changed, from a steady rain in 1889 to a storm of snow mixed with rain in 1893. Despite the inclement weather, President Cleveland stood for five hours to review the parade on snow-covered streets. The *New York Tribune* described the weather scene:

> Big wet snowflakes obscured the atmosphere and made it difficult to distinguish at a distance greater than a city square. Under foot it was worse yet. Slush and sleet struggled for supremacy with the fast falling snow, and to add to the discomfort a strong and cold wind was blowing. As the morning wore on it became positively dangerous to walk on account of the slippery condition of the streets ... In the early afternoon, it is true, the snow flurries ceased, but the wind grew steadily in energy and fitfulness. The sun tried hard about four o'clock to pierce the mist and gloom, but with little success.

Rain mixed with snow fell during the morning and until 12:50 P.M. The temperature rose to an even 32°F (0°C) at noon. A half-inch of snow covered the ground at the evening observation at the new United States Weather Bureau of the Department of Agriculture. Strong northwest winds gusted across the capital city, reaching a maximum of 29 mi/h (47 km/h) during the day.

William McKinley — 1897 — Clear and Mild

William McKinley's inauguration was the first to be captured and preserved on motion pictures, and also the first to have his address recorded live on a disc that could be played throughout the country.

The weather certainly cooperated in making both these innovations successful. Clear skies prevailed all day, with the temperature rising to 47°F (8°C) in midafternoon. Light winds averaging only 7 mi/h (11 km/h) added to the comfort of the day.

> Even the fickle and uncertain goddess of the weather put on her fairest smile to welcome the coming of "Prosperity's Advance Agent" [commented a reporter of the *New York Tribune*]. For twenty-four hours past this most capricious of months has ceased to bluster, and the purest azure skies and the mildest of western zephyrs has for the second time in a generation almost justified the

whimsical judgment which selected March 4 as the date for the greatest of open-air holidays at the National Capital.

William McKinley — 1901 —
Overcast, then Rain, Mild

President McKinley was returned to office by again beating his Democratic opponent, William Jennings Bryan. The margin was slightly larger than four years before. But the weather gods did not seem to favor him this time. A photograph of the inaugural address shows a number of umbrellas in the background, and needed they were.

> The ceremony of administering the oath of office to the President was marred by a slight fall of rain, which afterwards for an hour or more turned into a steady downpour. The success of the inaugural programme was not seriously clouded, however, by this disagreeable incident. The shower was an April one, and soon gave way to clearing skies, which assured the uninterrupted continuance of the outdoor celebration. Little discomfort was experienced by either the bodies which marched in the parade or the tens of thousands of spectators who witnessed it, and the day on the whole proved an unusually favorable one for inauguration purposes,

reported the *New York Tribune*.

Cloudy skies prevailed all day. Light rain began just as the swearing-in was taking place at 1:20 P.M. and continued to 3:45 P.M. The temperature rose to a maximum of 51°F (11°C) in the afternoon, up from 47°F (8°C) at noon. Light north and northwest winds prevailed, with a maximum speed of only 11 mi/h (18 km/h) in the afternoon.

Theodore Roosevelt — 1901 —
Clear Midday, Seasonable

At forty-two years of age, Theodore Roosevelt became the youngest man ever to serve as President of the United States. He succeeded to the office on September 14, 1901, after the tragic death of President McKinley, who had been struck by an assassin's bullet eight days earlier, while attending the Pan-American Exposi-

McKinley takes the oath of office in 1901. Courtesy of the Library of Congress.

tion in Buffalo, New York, and lingered on his deathbed for that time.

Vice President Roosevelt had been enjoying a camping trip on the slopes of Mt. Marcy in the Adirondacks of New York when a messenger delivered the sad news late on September 6. Roosevelt left immediately for the nearest railroad station and arrived in Buffalo early the next afternoon.

The local weather report for the day of his swearing-in was provided by the daily Weather Bureau summary: "The skies were clear from midmorning through midafternoon with light and variable winds. The high for the day was 76°F (24°C) and the morning low 59°F (15°C). No precipitation was reported and the sunshine for the day was 51 percent of the possible."

Theodore Roosevelt — 1905 — Sunny and Clear

Photographs of the Roosevelt inaugural procession proceeding to the Capitol show bright sunshine and sharp shadows of the men and horses, and flags were fully extended by the strong breeze. The day opened partly cloudy, but the clouds were thin and permitted 77 percent of the possible sunshine to penetrate during the

*Theodore Roosevelt rides in bright sunshine to the Capitol in
1905. Courtesy of the Library of Congress.*

day. About noon, the clouds began to disappear and the sky became completely clear after 1:00 P.M. At noon the thermometer
read a moderate 46°F (8°C), and the wind blew at 21 mi/h (34
km/h) from the northwest; it hit a maximum of 27 mi/h (43
km/h) in midafternoon. Patches of snow covered the ground here
and there.

The *New York Tribune,* a staunch supporter of the President,
called it "ROOSEVELT WEATHER," going on to comment a bit about
inauguration history:

> The inaugural parade, the most popular as well as the most bril-
> liant spectacle of the day's celebration, depends largely for its spec-
> tacular and inspiring effect on the weather conditions. March days,
> even in this favored climate, are subject to many sudden caprices,
> either balmy in their grateful mildness or chilling in their repro-
> duction of Arctic terrors, and dispensing indiscriminately cheer or
> gloom, according to their own sweet pleasure, but today "Roose-
> velt weather" prevailed. The morning hours were sultry, with the
> atmosphere heavy with moisture, and ominous clouds which
> threatened a genuine downpour gathered, causing a momentary

Clearing the reviewing stand area of snow in front of the White House for Taft's inauguration in 1909. Courtesy of the Library of Congress.

uneasiness among the assembled thousands, but as the hour approached for the President to take the oath of office, they were put to rout by the sun's genial rays, which were sufficiently strong to temper the chill March air . . . It became ideal inaugural weather, brisk and bracing, strikingly similar to that attending Cleveland's installation in 1885 and McKinley's entry into power in 1897. In the last half century a clear, bright, genial Inauguration Day has been the exception rather than the rule, and many have been hopelessly marred by the severity of the cold or the violence of the storms which inflicted the keenest suffering on those who participated in the celebration.

William H. Taft — 1909 —
After Storm, Snow Underfoot, Cold

William Howard Taft was the biggest man ever to be elevated to the presidency and he experienced the biggest snowstorm that ever fell on inauguration day. "Inaugurated under circumstances unique and unrivaled in the history of American statecraft, and

A snowy trip for Roosevelt and Taft in 1909. Courtesy of the Library of Congress.

weather conditions most unpropitious, William Howard Taft of Ohio, manfully and fearlessly gripped the sceptre of power of this nation," commented the magniloquent editor of the *Washington Post.*

Despite the prediction that a storm center from the Ohio Valley would pass rapidly eastward and be well out to sea by midnight, the center slowed its eastward movement, intensified, and wrapped the District of Columbia area in a severe storm. Moderate to heavy snow started about 9:00 P.M. on March 3 and continued until 5:00 A.M. the next day, though light flurries continued to filter down until 12:20 P.M. A strong northwest wind, peaking at 26 mi/h (42 km/h) at 8:12 A.M., blew the snow into mountainous drifts. The wind continued out of the northwest at close to 21 mi/h (34 km/h) all afternoon. The temperature rose to 32°F (0°C) by noon, when the sunshine recorder indicated six-tenths of the possible sunshine coming through the thinning clouds. With the thermometer rising to 35°F (2°C) in midafternoon, the snow turned to slush underfoot and the morning's 9.8

inches of snow melted down to 4.5 inches by the observation at 8:00 P.M. Skies became clear overhead for the evening festivities.

The *Washington Post* continued its description of the events that day:

> The seat of government of a celebrated nation was locked in the teeth of a furious blizzard, practically isolating it from the remainder of the world, and preventing the word of gladsome tidings from being spread abroad. Unable to make himself heard by the clamoring patriotic thousands assembled on the plaza of the Capitol, and loath to risk the health of the aged chief justice of the Supreme Court of the United States, who was to administer the oath of office, because of the snowstorm, howling winds, and low temperature, Mr. Taft obligated himself to honor and protect the American Constitution and flag in the Senate chamber, before an august body of men foremost in the councils of the nation.

Woodrow Wilson — 1913 — Overcast but Mild

Woodrow Wilson rode into office with a vast majority of the electoral vote, but with a minority of the popular vote. A bitter dispute had split the Republican Party between the ebullient Roosevelt and the easy-going Taft. The Virginia-born Wilson was the first Southerner to win election to the presidency since James K. Polk of Tennessee in 1845. And he was the first Democrat inaugurated in twenty years.

> The weather today is mild and at times the sun has shone with warmth that was uncomfortable for men marching under overcoats [observed the *Evening Star*]. The day dawned cold and gray, a solid bank of clouds completely shutting out the sun, threatening to carry out the official prediction of "unsettled weather in the late afternoon or tonight." A light south wind has fanned the city, however, bringing warmth, despite the thermometer, which has been between 45 and 50 degrees.

The official Weather Bureau summary of the day read: "day was pleasant, but overcast [enough] to be called cloudy. Light rain in early evening was hardly noticed." The temperature rose to 52°F (11°C) at noon and continued rising to a maximum of 59°F (15°C) in midafternoon. Light winds came from the south. Just a trace of rain fell between 7:15 and 7:20 P.M.

Woodrow Wilson — 1917 — Partly Cloudy, Cool, Windy

For the first time in the nation's history, the traditional inauguration day was held on a Sunday in 1917. It was a time of international crisis, for the United States had severed diplomatic relations with Germany. The nation should not be without a President for a day, it was decided, so Wilson quietly slipped to the Capitol on Sunday and there, in the President's room, the constitutional oath was privately administered by Chief Justice White.

Though the day started ominously, conditions improved rapidly by late morning. The *New York Tribune* described the scene:

> The sky was gray and wintry until noon. The Capitol loomed white against the threatening clouds, with the Stars and Stripes fluttering from the dome . . . The wind, which had been blowing freshly all morning, finally cleared an open space in the heavens, and the bright sun streamed down on the Capitol steps to a little white platform in the middle of the steps.

According to the official observer, raindrops ended at 4:50 A.M., though light drizzle continued until 8:55 A.M., and the sun appeared at 9:10 A.M. These were the first rays of sunshine that Washington had seen since the beginning of the month. Skies were partly cloudy at noon with the thermometer at 38°F (3°C). The wind came out of the northwest at 16 mi/h (26 km/h), gusting as high as 29 mi/h (47 km/h) in the afternoon.

Warren Harding — 1921 — Clear, Cool, Sunny

"Under a brilliant sky and in a keen atmosphere that had more than ordinary tang to it," the ailing President Wilson and President-elect Warren Harding drove from the White House to the Capitol in an automobile, the first time that other than horse and carriage had been employed, if we except some of the earlier horseback riders. "The cool breeze blew across the plaza snapping the flags in the brightest sunshine," continued the description in the *New York Times*.

President Harding breaks with tradition and drives in a Packard Twin Six on a sunny inauguration day in 1921. Photo no. 111-SC-73502 in the National Archives.

The official weatherman observed clear skies and 100 percent of possible sunshine prevailing through the day. The thermometer read 35°F (2°C) at noon on its way up to a maximum of 44°F (7°C) in the afternoon. The noon wind came out of the northwest at 12 mi/h (19 km/h), the peak for the day.

The *New York Herald* reporter added some personality to his account:

> They walked in a gorgeous morning, for Mr. Harding's famous weather luck had held. Either that or the Chief of the Weather Bureau is an utterly beloved Republican. Both made good. The new President told his friends laughingly in the New Willard late last night that he was not at all concerned about the weather, though the outlook was none too good at that hour, because he had a conviction that the sun would shine, that the singular good fortune that had clung to him since his nomination until his election would not desert him when he needed it most.
>
> Also Marvin, the weather man, stuck to his prediction of "clear and cold," and clear and cold it was. A savage northwest wind, which arose in the night, coming from regions where grapefruit do not flourish, swept away the clouds as a housewife clears the cobwebs from her ceiling, drove the dampness from the air, refined it, gave it ozone and left the rest for the sun to do.

Inauguration Day Weather

Calvin Coolidge — 1923 — Dark Outside

A modest farmhouse tucked away in a remote valley of the Green Mountains of Vermont was the site of Coolidge's inaugural. Just before midnight of August 2, 1923, the news of the death of President Harding in San Francisco earlier that evening reached Vice President Calvin Coolidge, who was vacationing at his family home in Plymouth. After consulting with the Attorney General in Washington, John Coolidge, father of the new President and a local justice of the peace, administered the oath of office at 2:47 A.M. on August 3.

Typical Green Mountain summer weather prevailed overnight. After a clear sunset, the sun rose with five-tenths cloud cover, according to the records of the nearest first-class weather station, at Northfield, some 45 miles north of Plymouth. Air movement was very light from the southwest, averaging only 1 mi/h (1.6 km/h), or practically a calm. The temperature during the early morning hours of the third dropped to a sunrise low of 63°F (17°C). With no weather fronts in the vicinity, conditions at Plymouth and Northfield would be close to identical.

Calvin Coolidge — 1925 — Mostly Sunny

A smiling Calvin Coolidge was photographed in top hat driving to the Capitol on a sunny noontime. The inauguration marked a milestone in the history of inaugurations. For the first time, the oath of office was given by a former President, William Howard Taft, who was now Chief Justice. And the words of the ceremony were carried to the far corners of the nation by means of radio phone, which had become a means of communication within the past four years.

The sun also smiled as the *New York Times* man described the scene:

> Brilliant sunshine, a breeze with a nip and a tang in it, made the weather conditions all that could be desired for the comfort of the participants in the inaugural ceremony and the thousands who stood in the Capitol plaza or along the route of the Presidential

Sixteenth Field Artillery in Hoover's rainy inaugural parade in 1929. Photo no. 111-SC-91420 in the National Archives.

progress . . . When the President left the White House it looked as if rain were imminent. The clouds and wind were threatening, but before it was time for Mr. Coolidge to take the oath, the sun broke through, the wind dispelled the clouds, and the President kept up his record of always taking office under a smiling sun.

The official data indicated that partly cloudy skies in the forenoon gave way to clear conditions again by 1:00 P.M. The sunshine count was 75 percent of the possible. The thermometer at noon read 41°F (5°C) before warming to an afternoon maximum of 52°F (11°C). Light southerly winds prevailed.

Herbert Hoover — 1929 — Sunny, then Rain

The thirty-first President of the United States, Herbert Hoover, was the first to be born west of the Mississippi River. West Branch, Iowa, was his birthplace, though he moved to California at an early age. His inaugural address was the first to be recorded on talking movies. Though the presidential party left the White House with the sun shining, light rain began to fall about noon, well before the oath of office was administered. The temperature at noontime stood at 48°F (9°C). The total rainfall for the day was a moderate 0.40 inch.

> It was bright at that hour [11:00 A.M.], with few indications of rain to come [observed the *New York Times*]. An hour later, however, it began to sprinkle, and soon there was a steady drizzle . . . Fear that the rainfall which began before the incoming President appeared on the platform, might ruin the picture proved unfounded . . . Before a great crowd, soaked by a cold penetrating rain . . . Falling harder and harder, the rain assumed downpour proportions, as the inaugural address ended.

Franklin D. Roosevelt — 1933 — Cool and Gray

"The only thing we have to fear is fear itself" were the historic words of 50-year-old Franklin Delano Roosevelt at his first inauguration, during the worst economic depression the nation had ever experienced. It was a good thing that he was young for the office, since twelve years later he would be taking the same oath to begin an unprecedented fourth term. Though rain threatened, it held off and the ceremonies went forward under cloudy skies and a noontime temperature of 42°F (6°C).

According to the *Washington Post*,

> The elements, so whimsy, on this day were kind . . . There was no rain. Midwinter sun tried once to give a freshening gleam to banner, steel and brass, to lend a luster to the uniforms and bright accoutrement to gold-fringed standards, dappled flank of charger, but finally gave up to dull gray skies.

Franklin D. Roosevelt — 1937 — Heavy, Chilling Rain

> Not in recent times has a President of the United States been inaugurated under more adverse weather as far as spectators were concerned than was Mr. Roosevelt yesterday. Even the blizzard which marked William Howard Taft's [inauguration] in 1909 was

less biting and penetrating than the rain and cold which pelted Capitol Hill yesterday and caused thousands to shiver as they watched the Inaugural parade. Rain fell in torrents throughout the morning, leaving water several inches deep on sidewalks and streets and turning the lawns on Capitol Hill into muddy, clay pools. Its volume increased a few minutes before noon. Cold that penetrated several thicknesses of clothing left spectators with their teeth chattering.

So the *Washington Post* described the first inauguration to be held on January twentieth, in accord with the Twentieth Amendment.

The official weatherman noticed that some ice formed on trees and roofs of buildings during the morning, but none on the ground or pavements. The temperature at noon stood at 33°F (1°C) and climbed only to 36°F (2°C) in the afternoon. Light rain began in the early morning but reserved its heaviest fall for the period from 10:00 A.M. to 3:00 P.M., when the inaugural ceremonies were in full swing. The greatest hourly amount all day, 0.36 inch, fell during the hour from 12:00 to 1:00 P.M. In the five-hour period covering noon over one inch fell of the day's total fall of 1.77 inches. The wind averaged 10 mi/h (16 km/h) from the northeast during the noon hour.

Franklin D. Roosevelt — 1941 — Clear and Brisk

Franklin D. Roosevelt broke all precedents when he took the oath of office for the third time on January 20, 1941. The weather elements smiled on the occasion, as the *New York Times* reported:

In the bright cold sunshine . . . most of the people were stamping their feet and rubbing their hands in efforts to keep warm in the cold wind that whistled about with the temperature in the twenties . . . The sun shone with a midsummer, dazzling glare, which the cold turned to a sham.

The official record agreed. Clear skies and sunshine were featured that day. With a cloudless sky at noon, the temperature reached 29°F (−2°C) and the wind blew from the northwest at 15 mi/h (24 km/h). These auspicious conditions continued through the day, which rated 89 percent of possible sunshine.

Franklin D. Roosevelt — 1945 — Cloudy, Snow on Ground

Precedent again was broken when a failing President Roosevelt took the oath of office on the south portico of the White House. All things combined to require a simple ceremony: the existence of a global war, the President's ill-health, and the weather.

> An overcast sky parted and a patch of blue appeared overhead just as Franklin Roosevelt took his position at the speakers' stand on the south portico of the White House at noon today for his fourth inauguration as Chief Magistrate of the country . . . An overnight storm had left a light fall of snow on the ground and on the leaves of the nearby magnolia trees. The thermometer stood at one degee above freezing but, unlike all others, the President stood without an overcoat,

reported the *New York Times*.

Cloudy skies prevailed in the morning and over the noon hour until 3:00 P.M., when breaks appeared in the overcast, and finally totally clear conditions came at 7:00 P.M. Light snow fell until 8:13 A.M., whitening the ground, but only a trace remained at 7:00 P.M. The thermometer at noon read 35°F (2°C). The sunshine during the day totaled 23 percent of the possible. Winds in early afternoon averaged 10 mi/h (16 km/h) from the north.

Harry S Truman — 1945 — Fair

Vice President Harry S Truman took the oath of office at 7:09 P.M. on April 12, 1945, in the Cabinet Room of the White House. Fair weather had prevailed all day as a result of a strong high-pressure area standing just off the Delmarva peninsula. The temperature mounted to a high of 74°F (23°C) that afternoon.

Harry S Truman — 1949 — Clear and Moderate

"The weather contributed materially to the brilliant spectacle. Above the White House, the skies were blue without a cloud. The clarity of the atmosphere made colors stand out everywhere. The day was cold and brisk," was the concise, but all-inclusive report

Clear with 1/10 cirrus: 38°F; winds light and variable for Harry Truman in 1949. Photo no. 306-NT-354A-14 in the National Archives.

in the *New York Times* on Harry Truman's second inauguration.

The Weather Bureau agreed. Skies with only wisps of high clouds favored the noontime hours. The thermometer stood at 38°F (3°C) at noon and rose to 43°F (6°C) at 2:00 P.M. for the maximum of the day. Light winds of about 10 mi/h (16 km/h) prevailed during the early afternoon. One hundred percent of possible sunshine was registered.

Dwight D. Eisenhower — 1953 — Sunny and Pleasant

As they had several times during World War II, the weather gods favored General Eisenhower as he stood to take the oath of office of the presidency of the United States. As the *New York Times* reported, "... with the sunny skies bringing mild temperatures, jovial good humor and a carnival spirit to the city."

The scattered high clouds observed by the Weather Bureau at noon were completely gone by 1:30 P.M. when the ceremonies were beginning, and the *Washington Post* added:

> By this time the Weather Bureau forecast had run into "Eisenhower luck" and had been routed. A large part of the sky was free of clouds. There was a haze but the sun was coming through to glint on the horns of the musicians, add vividness to the Flag, and give brightness to the drama under the Capitol dome . . . The day was exceedingly pleasant for Washington on January 20 — the temperature reached 54 at 3 P.M.

The sunshine register showed 100 percent of the possible.

Dwight D. Eisenhower — 1957— Breaking Clouds, Moderate

> Under a cloudy pall that just for an interval gave way to a blaze of sunlight . . . The radiance broke through as the Chief Executive stood on a Corinthian platform on the east steps of the Capitol . . . He was standing to the cheers of 25,000 out on the plaza, with arms outstretched, when a burst of sunlight pierced the gloom.
>
> What has come to be known as the "Eisenhower luck" held good as the morning mists disappeared and the January sun broke through slowly moving clouds to the South, complete with a burst of sunshine as he repeated the oath after Chief Justice Earl Warren.

The official weatherman was not as enthusiastic as the reporter of the *Washington Post*. Rain had ended at 9:26 A.M., but overcast skies continued until 1:30 P.M., when some breaks in the clouds appeared. A very light rain shower fell at 1:57 P.M. The temperature at noon stood at 44°F (7°C) and rose only one degree an hour later. Winds were light from the southeast.

John F. Kennedy — 1961 — Sunny and Chilly After Heavy Snow

For the first time since Taft's inaugural, a heavy snowstorm figured in the 1961 inauguration weather scene when a storm center from the Gulf of Mexico advanced over the southern Appalachians toward Washington on January 19. Flakes began to fall in the Washington area early on the afternoon and threatened to tie up all traffic on the eve of John F. Kennedy's inauguration. Seven inches had fallen by midnight, but the fast-moving storm passed by with little additional accumulation. Nevertheless, it took a herculean effort by city and military snow-removal equipment to make Pennsylvania Avenue ready for the inaugural procession.

The *New York Times* reporter caught the scene:

> A blanket of 7.7 inches of newly fallen snow, bitter winds, and a sub-freezing temperature of 22 degrees held down the crowds . . . A Siberian wind knifing down Pennsylvania Avenue in the wake of last night's snowfall . . . But the crowds swelled under a cheering, if not warming, sun from a cloudless sky . . . The sun was bright as the cars rolled down Pennsylvania Avenue, but the wind was sharp. Many in the crowd kept their hands in their pockets to keep warm, and the cheers sounded thin in the cold air.

After a heavy snowfall overnight, the flakes tapered off to a light snow with flurries continuing until 6:40 A.M. Eight inches lay on the ground at noontime. The clouds commenced to break at 8:00 A.M. and the skies became completely clear after 11:00 A.M. The rest of the day enjoyed brilliant sunshine. The thermometer read 22°F (-6°C) at noon and rose to a maximum of 26°F (-3°C) by 4:00 P.M. A northwest wind blew strong all day, peaking at 37 mi/h (60 km/h) at 10:27 A.M.

Lyndon B. Johnson — 1963 — Cool, Clear

The jet-age setting of the cabin of *Air Force One* provided the setting for the oath-taking of Vice President Lyndon Johnson on that tragic day in Dallas, November 22, 1963. At 2:29 P.M., less than two hours after the death of President Kennedy, the oath was spoken by Johnson with Mrs. Kennedy standing at his side. The plane was parked on the apron of a runway of Love Field on the outskirts of Dallas. At this time the temperature hovered close to the day's maximum of 68°F (20°C). The passage of a cold front in the morning left the skies clear and the wind moderate from the northwest.

Lyndon B. Johnson — 1965 — Cloudy, Pleasant

> For Inauguration Day the weather was surprisingly good, looked at against the record of such days. At noon, when Mr. Johnson was about to repeat the oath after Chief Justice Earl Warren, the Weather Bureau reported a temperature of 38 degrees, 18 degrees warmer than noon four years ago. The sun was trying to break through a haze ... The crisp weather and sunny skies drew hundreds of thousands to watch his return from the swearing-in on Capitol Hill ... The wind was out of the Southwest, from the direction of the Texas hill country, and it seemed to blow something into all that happened in the Inaugural drama of 1965

— as reported by the *Washington Post*. Mostly overcast skies prevailed during the noon hours, though the day as a whole had 58 percent of possible sunshine. At noon, clouds covered nine-tenths of the sky, and the temperature stood at 38°F (3°C), rising to 47°F (8°C) in the afternoon. Winds from the southwest averaged 12 mi/h (19 km/h).

Richard M. Nixon — 1969 — Overcast and Chill

> Physically, it was a day out of Edgar Allan Poe, dun and drear, with a chilling northwest wind that cut to the marrow, and a gray ugly overcast that turned the city the color of wet cement. No graves yawned and no lions roared in the streets, in the Shakespeare manner, but the gloom of the elements seemed to have infected most of the proceedings.

Russell Baker, the droll columnist of the *New York Times*, did not approve of what was going on around him.

Skies continued overcast throughout the day, with light rain starting at 4:30 P.M. Some sleet mixed with the rain briefly in the early evening. The temperature at noon stood at 35°F (2°C) and remained steady during the next two hours. Wind flow from the northeast measured about 11 mi/h (18 km/h) in the early afternoon.

Richard M. Nixon — 1973 — Raw, Chilly

The *Washington Post* chronicled the day in this way:

> About the only thing that went wrong with his day was that capricious element that plagues so many inaugurals, the weather. Instead of the mild day that had been forecast, it was raw, chilling weather that greeted the early arrivals at the Capitol . . . The west to northwest winds were whipping across the grandstands and the official presidential pavilion in gusts up to 30 miles an hour, causing people to stamp their feet in an effort to keep warm. Occasionally the sun broke through the thick, swiftly moving clouds, but for the most part it was a somber setting for the country's most formal ceremony.

Cloudiness prevailed most of the day with only 4 percent of the possible sunshine appearing. Noontime found the sky overcast with clouds at 4500 feet. The thermometer read 42°F (6°C) and rose only one degree in the next two hours. Wind came from the northwest varying from 16 to 24 mi/h (26 to 39 km/h). No precipitation fell.

Gerald Ford — 1973 — Rainy and Humid

President Nixon submitted his resignation on August 9, 1973, a day that "began in the mist and rain of a humid Washington morning," according to the *New York Times*. Soon afterwards, Vice President Gerald Ford took the oath of office in the historic East Room of the White House. The thick weather prevailing at Washington was caused by a stationary front that sprawled from southern New England to West Virginia. The temperature at the National Airport ranged from a maximum of 83°F (28°C) to a minimum of 70°F (21°C). Rainfall during the day amounted to 0.32 inch.

Jimmy Carter — 1977 — Cold, Sunny

Jimmy Carter broke all precedents of the inaugural parade by traveling on foot rather than on horseback, in a carriage, or in an automobile. The *New York Times* correspondent described the event: "He walked a mile and a half in subfreezing weather from the Capitol to the White House with his wife Rosalynn . . . while a sharp east wind whipped his gray hair . . . He took his wife's hand, descended from the stand and walked across the frozen lawn to the White House."

The *Washington Star* reported the prevailing weather:

> Squinting in the fresh sunshine . . . A chilled crowd estimated at 150,000, well bundled even though the sun was out and the temperature was moving up noticeably for the first time in days. And a large and shivering crowd concentrated mostly on the sunny side of Pennsylvania Avenue . . . [which] had been scoured free of the last ice, but many of the spectators had to stand on icy patches along the curbside.

The scattered high clouds of the thin cirrus type predominating the sky all day permitted 96 percent of possible sunshine. At noon there were four-tenths high cirrus, but they disappeared completely by 5:00 P.M. The temperature at noon stood at 28°F (−2°C) and rose to a high of 33°F (1°C) by 3:00 P.M. Winds averaged about 12 mi/h (19 km/h) during the early afternoon.

Ronald Reagan — 1981 — Cloudy, Mild

At 69 years of age, Ronald Reagan was the oldest man to assume the presidency and, after serving five months, was the oldest man ever to hold the office. Another of the firsts of the incoming administration was a sensible one: the decision to hold the inaugural ceremonies on the west portico of the Capitol instead of at the traditional location on the east portico. Because the Capitol building is oriented north and south, this opened the inaugural platform to the afternoon sun. As it turned out, temperatures remained in the comfortable 50s all afternoon.

"The clouds that covered the sun at dawn moved south during the morning, and the winter sun broke through in the inaugural ceremony, sending the temperature to 56 degrees and making it one of the warmest inaugural days on record," reported the *New York Times*.

The *Washington Post* commented on the weather in its special section on the inauguration: "In the distance, obscured by a slight haze on an otherwise perfect, balmy day . . . The sun retreated behind darkening clouds yesterday afternoon, the rain that was expected didn't fall."

The official thermometer at National Airport ranged from a low of 38°F (3°C) to an afternoon high of 56°F (13°C). The sky from sunrise to sunset was overcast. A little sunshine did get through from time to time, the recorder giving 14 percent of the possible. Wind averaged 9.1 mi/h (14 km/h) with the highest gust at 15 mi/h (24 km/h). A trace of precipitation fell during the evening hours.

CHAPTER FOUR

The World of Sport

Sporting events are especially subject to the vagaries of the weather, because most contests over the years have been staged outdoors where rain, wind, or cold can make playing conditions difficult or impossible.

Probably rain has caused more headaches for the national pastime of baseball than for any other sport because all the scheduled games have to be completed, and this often means accumulating double-headers toward the end of the season, the bane of any manager's planning. The outcome of a World Series has been influenced by postponements that permitted an ace pitcher to have an extra day's rest between starts on the mound.

Football is customarily played on the scheduled day regardless of rain or snow, but the playing surfaces have often been transformed into a slippery sheath of mud or ice. This greatly upsets the quality of play and brings chance to the fore as a determining factor in the outcome of such contests.

Record possibilities often vanish when an encounter with adverse weather such as a heavy or muddy track at the Kentucky Derby eliminates the prospect of fast times. Unseasonable heat has sapped the energies of runners in the Boston Marathon and slowed the race.

Golf is especially affected by weather changes that result in unequal competing conditions, for example, when the wind suddenly increases during a day's play, or an afternoon thundershower provides late finishers with wet fairways and greens after earlier players have enjoyed normal surface conditions. As a result of several recent misfortunes, lightning has become a menace to orderly completion of play in some major tournaments.

The Winter Olympic Games must have suitable weather before and during the contests. A near-disaster was narrowly averted at Lake Placid in 1932 when an unseasonably mild January threatened to remove the snow cover entirely, before a fortunate change in the weather brought a cold front and a snowfall barely sufficient for the games to finish, though they were two days late in so doing.

Numerous efforts of late have been made to circumvent the effects of adverse weather, especially in professional sports with their large economic involvement in a single contest. Superdomes have been constructed to move major sports indoors. Artificial surfaces have been developed to permit play on damp fields. Quick-drying tennis courts have allowed resumption of play soon after a rain ceases. Snowmaking machines have provided surfaces for skiing when natural snow is absent. Fields, tracks, and golf courses have been constructed with good drainage and drying foremost in mind.

Yet the weather still remains an important factor in all outdoor sports. Contestants always hope for ideal conditions for their particular sport so that they can perform at their best.

Baseball

The World Series

Timely Postponements in 1903

Two postponements at convenient times during the first World Series, in 1903, enabled "Deacon" Phillippe of the Pittsburgh Pirates and Rural Retreat, Virginia, to pitch five complete games in a single series against the Boston Pilgrims of the American League. The series then ran to the best five out of nine games. Phillippe pitched 44 innings, while the rest of the Pirates' staff saw duty in only 26 innings. The Pirates' stalwart won three and

Early game at Elysian Fields in Hoboken, New Jersey. Courtesy of the Library of Congress.

lost two games, but his teammates lost the series by five games to three, as a result of their failure to win any games when Phillippe was not on the mound. The latter pitched his five complete games on October 1, 3, 6, 10, and 13. The rain postponements came on October 5, 9, and 12, and October 4 and 11 were Sundays with no play.

The Shutout Series of 1905

There was no series in 1904 between the National League and American League champions, but in the 1905 match a rain delay enabled young Christy Mathewson of the New York Giants to pitch and win three shutouts in the five-game series. Joe McGinnity pitched the other two games for the Giants against the Philadelphia Athletics. In those days of iron men, each team employed only three pitchers in the five games played. Each game was won by a shutout, with the Giants taking four to the Athletics' one. Mathewson pitched his consecutive shutouts on October 9, 12, and 14. The eleventh was a rain day.

The Cold, Snowy Series of 1906

"At 1 o'clock a slight drizzle began and at times flakes of snow fell" on opening day of the intracity World Series between the Chicago Cubs and White Sox on October 9, 1906. The maximum temperature of the day, 46°F (8°C), came early in the morning, and it tumbled all day toward the freezing mark.

Game No. 2 was played under the most frigid conditions of any World Series game. The temperature ranged that day from a low of 30°F (−1°C) to a high of only 39°F (4°C). The cold weather cut the attendance to 12,693, down from an expected 20,000. The conditions for Game No. 2 were described by the Associated Press dispatch:

> The weather was bitterly cold, the mercury hovering at or below the freezing point all afternoon. As was the case yesterday, a few flakes of snow fell, but they were not needed to add to the general discomfort of the spectators and players. Between nine and ten thousand enthusiasts, bundled from head to foot, braved the rigors of the weather, but they by no means filled all the seats.
>
> With the frost nipping fingers and toes, perfect baseball was an impossibility. The practice of both teams was exceedingly ragged, the cold hands of the infielders refusing to cling to the ball, while the outfielders missed many flies which were blown out of reach by the wind. In actual play the form was better.

It was still cold for game No. 3 on October 11, with a maximum of 43°F (6°C), up from a morning low of 27°F (−3°C). It warmed up for Game No. 4 to 64°F (18°C) and for Game No. 5 was a balmy 74°F (23°C). The White Sox proved better polar bears than the Cubs by winning the series four games to two.

Rain Delays — 1911

The longest rain postponement in World Series history occurred in 1911 when the playing of Game No. 4 between the Philadelphia Athletics and the New York Giants was delayed for a full week. It was scheduled for Wednesday, October 18, at Shibe Park in Philadelphia, but was not played until Tuesday, October 24.

On the eighteenth, a moderate rain of 0.75 inch fell at Philadelphia and additional amounts on each day following added up to a six-day total of 2.89 inches. Newspaper reporters dubbed the field "Shibe Pond." During the successive delays the Giants returned overnight to their New York homes or hotels. The New York weather bureau received a record number of phone calls from impatient fans. The long layoff enabled Christy Mathewson, now in his tenth season, to pitch successive games, but he lost both to the Athletics, who took the series four games to two. The series required thirteen days to complete.

Darkness in 1912

The only World Series game that was started and did not reach a decision on the same day occurred on October 9, 1912, when Game No. 2 was called on account of darkness, after eleven innings of play. Each team scored a run in the tenth inning, but was held scoreless in the eleventh. The game ended at 6–6 when Umpire O'Loughlin called it. The results were included in the composite statistics of the series, but Game No. 2 was replayed the next day, ultimately extending the series to eight full games.

The length of the called game was only two hours and thirty-eight minutes, but the standard time in effect had the sun setting at 5:13 P.M. because Boston lies in the eastern part of the time zone.

Cloudiness increased the gathering gloom. "A cold southwest wind blew over the top of the grand stand and made heavy coats and even furs, things of joy," reported the *New York Herald.* "It was cloudy all day, except for one or two brief intervals when the sunlight struggled through." The game had been scheduled to begin at 2:00 P.M., but was delayed by the opening day ceremonies presided over by Mayor "Honey Fitz" Fitzgerald, John F. Kennedy's maternal grandfather.

"Darkness was fast descending when the Red Sox came up for their turn at bat [in the eleventh inning], and O'Loughlin held a hurried consultation with Rigler. The latter was seen to shake his head in the negative and the game continued." The Red Sox went down quickly with three infield groundouts, and the game was called, the only such instance in World Series history.

Stormy 1962

The 1962 series between the New York Yankees and the San Francisco Giants stands second in the number of postponements and is the only series that had games postponed in each home city. Game No. 5, scheduled for Yankee Stadium on October 9, was put over a day by a rainfall of 0.23 inch. When the teams tried to play Game No. 6 at San Francisco on Friday, October 12, a three-day delay ensued. What was described as "the worst early autumn storm since 1904" struck the Bay Area. The storm, lasting from the tenth to the thirteenth, dropped the large total of 7.29 inches. It was not until Monday, October 15, that the field dried out sufficiently to resume play. The delay enabled Ralph Terry to pitch a strong four-hit shutout in Game No. 7 and win the series for the Yankees.

The Rain-outs of the 1970s

The 1970s suffered more from rain interference with World Series games than any other decade. There were two postponements in the 1950s and only one in the series in the 1960s, but from 1971 through 1980 there were four series with rain delays.

The second game in 1971, on October 10, between the Baltimore Birds and the Pittsburgh Pirates, was washed out when a heavy downpour of 2.64 inches filled the Baltimore rain gauge and spoiled the anticipated Sabbath pleasure of millions of TV watchers.

On October 17, 1972, Game No. 3 between the Cincinnati Reds and the Oakland Athletics was put over a day. A week of rain in "sunny" California, climaxed with 0.67 inch on the sixteenth and 0.09 inch falling at Oakland on the seventeenth, made the field unplayable.

The sixth game of the memorable 1975 series between the Cincinnati Reds and the Boston Red Sox on October 18 was washed out when a deluge of 1.97 inches descended within 24 hours. This was part of a five-day northeaster from the sixteenth to the twentieth in which 2.81 inches fell. The soggy ground did not permit play until the twenty-first, after a three-day delay.

Heavy rains on October 20, 1976, at New York City caused

the postponement of the fourth game in the four-game series between the Cincinnati Reds and the New York Yankees. The Central Park rain gauge measured a total of 1.68 inches.

Rain began to fall three hours before the scheduled evening opening game of the 1979 World Series at Baltimore on October 9 and caused the first postponement of an opener in the history of the classic. Around noon the next day the rain turned to snow and fell heavily for several hours, though mostly melting on the ground. But the evening game went on as scheduled despite wet grounds and a gametime temperature of 41°F (5°C). Baseball writers accused Commissioner Bowie Kuhn of having outfitted himself with thermal underwear. He wore no overcoat to demonstrate that his decision to play ball that evening was the right one.

Indoor Rain-out
No, it did not rain in the Astrodome on June 15, 1976, but it rained so hard in the surrounding area of Houston that the streets and grounds around the covered ballpark were flooded. The Astrodome was near the center of a deluge that dropped 10.12 inches into the nearest official rain gauge at Westbury.

So the Astros and the Pittsburgh Pirates had a "night off." However, they were stranded at the Astrodome along with about 20 employees. The umpires and most of the Astrodome staff were unable to reach the park because of the flooded street conditions.

It was the first such postponement in the first eleven years of the Astrodome. A spokesman said: "It's not exactly a rain-out, it's a rain-in. We're bone dry inside."

All-Star Games

Commencing in 1933, the series of annual All-Star games have been especially fortunate in their attendant weather. Of the fifty-two games played through 1983, only two have had a rain interference.

1952
The game at Philadelphia on July 8, 1952, was called at the end of five innings when "a drizzling rain that had been sprinkling a

crowd of 37,785 from the start cracked down in earnest," according to the *New York Times* report. After a 56-minute wait, the umpires called the game. A total of 0.53 inch fell that day. The National League was declared the winner of the contest by a score of 3–2.

1961

There were two All-Star contests in 1961 and each had a weather experience. On July 11, "the wind at Candlestick Park almost blew the wick out," wrote a reporter of the game at San Francisco, which the National League won by 5–4 in ten innings. The second game was played at Boston on July 31. At the end of nine innings, with the score tied 1–1, an east wind brought a Massachusetts Bay drizzle that halted the contest. After a wait of only 21 minutes, the weatherwise umpires called the game, the only tie in the history of the series. The airport weather station recorded 0.25 inch, and heavier amounts fell to the south of the city.

1966: The Hottest

"The game was a sizzler. So was the weather. It was the sort of day on which sensible men would have completed their chores as expeditiously as possible in order to escape from the inferno of Busch Memorial Stadium and seek hasty relief in the comfort of their air-conditioned indoors," wrote Arthur Daley of the *New York Times*.

The hottest of All-Star games was played in downtown St. Louis on July 12, 1966. A record heat wave was in progress with the thermometer topping 100°F (38°C) on six consecutive days. On the afternoon of the game it reached 104°F (40°C) in the downtown section and 105°F (41°C) at the airport. These figures should never be exceeded since the All-Star games are now played at night.

Despite the temperature, it was a well-played game. "But the heat, abnormal for this traditionally well broiled area, had no effect on the calibre of play," commented Leonard Koppett, regular sports writer for the *Times*. It went to ten innings, until Maury Wills singled to drive in the winning run, enabling the National League to triumph over the American League by 2–1.

Dodgers rained out in Los Angeles in 1976. National Baseball Hall of Fame.

"Dodger Weather"

"It never rains at Dodger Stadium," declared the Los Angeles baseball club's pressbook in boldface type for many years, and this was almost true. Since the baseball season coincides with California's dry season, the averages show that only 4 percent of the annual rainfall can be expected in the six months from May through October.

From the time that the Dodgers forsook Brooklyn and went west in 1958 through the conclusion of the 1975 season, only one game had been rained out. This occurred on April 21, 1967, during a five-day period when 1.68 inches of rain dropped on the city. But the seasons of 1976 and 1977 — strangely enough, years of extraordinary drought throughout California — each brought three postponements. The 1976 opener, on April 12, was washed out by a 0.33-inch rainstorm. Then tropical storm Kathleen spoiled a September weekend with very heavy rains: 1.74 inches on Friday the tenth and 0.30 inch on Saturday the eleventh, making consecutive double-headers necessary on Sunday and Monday. The double rain-out of September 1976 was repeated early in the next season, a second year of drought. A 2.02-inch rain fell on May 8 and 0.75 inch on the ninth, washing out consecutive games. A second tropical storm in two years — this time Doreen — brought very heavy rains to the Los Angeles area on August 17–18, 1977, with a 2.07-inch rainfall, marking the seventh postponement in the Dodgers' twenty-year history on the West Coast, a much better record than that of any other major league baseball team over that period.

Rain-related postponements didn't end in 1977, for a late-season storm on April 15, 1978, dropped 0.8 inch on downtown Los Angeles and led to the postponement of the game scheduled between the Dodgers and the Atlanta Braves, and on September 5, moisture carried north from Hurricane Norman, off the coast of Baja California, brought a 0.35-inch rain that postponed the game between the Dodgers and the San Francisco Giants.

After a four-year hiatus in rain-outs, two games were washed out in the 1983 season. The game between the Dodgers and the San Diego Padres was called after a wait of 69 minutes resulting from a 0.61-inch rain on April 10–11. And a game scheduled for August 18 between the Dodgers and the New York Mets was delayed for nearly three hours and finally rescheduled for later in the season because a tropical storm dropped 1.10 inches of rain on Dodger Stadium, bringing the eleventh postponement in the twenty-five years since the Dodgers left Brooklyn.

Opening Series Snow-out at Boston in 1933

Boston underwent a long siege of snow and rain during the middle of April 1933, which caused the postponement of the entire opening series of the season between the New York Giants and the

Boston Braves. Thus, the teams started the season with the prospects of four double-headers facing them. Boston was deluged with 1.32 inches on the twelfth and 1.46 inches on the thirteenth. Five inches of snow fell on the thirteenth, with the mercury ranging from 32°F (0°C) to 36°F (2°C). The game on the twelfth was rained out, on the thirteenth snowed out, and on the fourteenth and fifteenth postponed on account of wet grounds and threatening weather. The precipitation continued on every day through the nineteenth, though temperatures moderated considerably.

Opening Day Snowstorm at Boston: April 14, 1953

It was snowballs instead of baseballs today at Fenway Park. Members of the Red Sox offered "Merry Christmas" greetings instead of wishing each other good luck for the opening of the American League season as an unexpected snow storm forced postponement of today's and tomorrow's games between Washington and Boston.

Several of the players tried out skis, strapped onto their spiked baseball shoes, and a couple who had never seen snow before even tasted the stuff.

*　*　*

Mel Parnell, who was to have pitched for the Sox today, tried his luck with snowballs, hurling from skis.

It was Parnell's first snowball and his first try on slats. Mel is from New Orleans, and while he had seen snow before, he never had made or tossed a snowball.

The ball players even batted a few snowballs, which, of course, shattered when struck, to the accompaniment of loud laughs.

*　*　*

It was the first snowed-out ball game since 1933, when a Braves-Giants series was called because of snow.

— *New York Times*, April 15, 1953

A northeaster dropped 2.17 inches of precipitation on Boston on April 13–14, and ended up with 2.2 inches of snow on the ground at Boston Airport. The temperature on the fourteenth ranged from a low of 32°F (0°C) to a maximum of only 39°F (4°C). It warmed up to 58°F (14°C) the next afternoon, but there was still snow on the field in the morning so the game was called on the

Mel Parnell of the Red Sox pitches snowballs instead of baseballs on the scheduled opening day at Boston, 1953. Courtesy of Weatherwise.

fifteenth also. Snow also caused postponements at St. Louis on the eighteenth and at Pittsburgh on the twentieth of that chilly spring month.

The Opening Game Snow-out of 1982

The New York Yankees were scheduled to open the season against the Texas Rangers on Tuesday, April 6, but instead opened the season six days later on Sunday, April 11, by losing a double-header to the Chicago White Sox. This is probably the first time that any team opened the season with a double-header. The reason was a late-season snowstorm which covered most of the Midwest and all of the Northeast. "Twelve inches," Esposito, the Yankee superintendent, reported at midafternoon of the sixth after he had measured the deepening snow against the boots that he reluctantly put on for each outdoor expedition. "Three and a half feet where the wind drifts it," he added.

The New York Mets, too, were snowbound at Shea Stadium, where the bus to take them to Philadelphia to open against the Phils was snowed in. When they did play on Thursday night, only 15,345 attended. There were 50,000 empty seats at Veterans Stadium on account of the inclement weather.

Other cities which had postponements on account of snow and cold were Milwaukee, Chicago, Detroit, Pittsburgh, and Baltimore.

Weather Disruptions in April 1983

After a mild winter in the Midwest and Northeast, the weather turned unfavorable when the baseball season opened and grew progressively worse through the second week of scheduled play. On Sunday, April 17, the New York Mets game at St. Louis was called off by a "freakish snowstorm." Other games were played in just above freezing temperatures: Detroit at Chicago 34°F (1°C) and Kansas City at Milwaukee 36°F (2°C). With a light schedule on Monday, no games were postponed; but an all-day snowstorm on Tuesday, April 19, disrupted play over a wide area of the Northeast, and an out-of-season rainstorm on the Pacific Coast caused unusual rainouts:

Pittsburgh at New York	snow
Chicago at Philadelphia	cold
St. Louis at Montreal	snow
Kansas City at Detroit	cold
Oakland at Anaheim	rain
Los Angeles at San Francisco	rain

Other games that day were played in nipping temperatures: New York Yankees at Chicago 36°F (2°C), Texas at Baltimore 40°F (4°C), and Cleveland at Toronto 34°F (1°C). High winds prevailed at Baltimore and Toronto to add to the wind chill.

On Wednesday, April 20, cold conditions still plagued the Northeast and more rain soaked the Far West. Games were played at Philadelphia in 41°F (5°C) weather, at Boston in 45°F (7°C), and at New York in 40°F (4°C) with only 4041 fans attending. Rain or wet grounds again postponed contests at San Diego and Anaheim.

Football

The First "Football" Game — 1869

The first intercollegiate "football" game played in America brought Princeton and Rutgers together at New Brunswick, New Jersey, on November 6, 1869. The rival captains agreed upon a set of rules that varied slightly from those adopted by the London Football Association, which essentially concerned soccer. The round ball could not be carried, but could be kicked or butted with the head. There were twenty-five players to the side. The posts supporting the crossbar of the goal were placed twenty-five yards apart, making a wide goal over which the ball must be kicked to score. The first team to score six goals was declared the winner. About 250 rooters, including 50 from Princeton, witnessed the classic match.

The day was cloudy, according to the local observation by P. Vanderbilt Spader, who had kept weather records at New Brunswick since 1847. The wind blew from the northwest and kept the maximum temperature to only 48°F (9°C). Rutgers won six goals to four. In a return match a week later, when Princeton was allowed to use its special free kick and different rules, the Tigers won by eight goals to none for Rutgers, the beginning of a mastery over their New Jersey rival that lasted until 1938, though there were long lapses in the series.

First Harvard-Yale Game in 1875

After a conference between student representatives from several colleges, it was decided to draw up a set of rules that would include features of both rugby and soccer. The new code was called the "Concessionary" rules, which were more rugby than soccer. They permitted one to run with the ball and to tackle, but only above the waist. A round ball was employed and only kicked goals counted in the scoring. If one made a touchdown by rushing across the goal, he was entitled to a free kick at the goal for a point.

Yale and Harvard came together for their first historic match on November 13, 1875, at Hamilton Park, New Haven. "The afternoon was fair for the sport," reported the *New York Herald*. Another reporter present said it was "a cold and cloudy day." Two thousand spectators, including many ladies, attended. Harvard won by four goals to none.

Two Princeton "scouts" attended the game and were favorably impressed by the new rules and the style of play. They convinced the teams on their schedule in the New York area to adopt the new game. The 1875 season marked the last in which soccer was a major sport, for the transition from soccer, to soccer-rugby, to football was well under way.

The First Rose Bowl Game: January 1, 1902

At the end of the undefeated season of 1901, Coach Fielding H. (Hurry-up) Yost took his thirteen Michigan football players to Pasadena to play in the first football game sponsored by the Tournament of Roses Association. It was a lopsided game with Michigan rolling up 49 points to 0 for Stanford, a team that "Hurry-up" had coached the year before. The game was cut short by six minutes to end the "track meet." No more Rose Bowl games were played until 1916, when the West felt confident they could field a team to match their Eastern rivals.

Fortunately, the *New York Times* gave good coverage to the initial game in this then-remote section of the country, and, luckily for us, the account ended with an informative paragraph: "The day was perfect, though possibly a trifle warm for football. The field upon which the teams met was as smooth as a floor, but very dusty. The crowd numbered 7,000."

No doubt the field was extremely dry and dusty, for California was going through one of its driest early winters. No rain had fallen at Los Angeles or Pomona since November 10. The maximum temperature at each place on the afternoon of New Year's Day 1902 was 61°F (16°C).

Yale Outsloshes Harvard: November 24, 1923

In the eyes of the students and alumni of Harvard and Yale, the concluding contest each season between the two schools is "The Game." Since the first contest, played in 1875, the rivals had met each season and had been favored by mostly fair weather for their pigskin confrontations. But this all ended on the sodden afternoon of November 24, 1923, when Raymond "Ducky" Pond, the aptly named Yale halfback, sluiced up a fumble and sloshed to the first touchdown scored by Yale at Cambridge since 1916, "in a rain that soaked them to the skin and turned the gridiron into a cross between a quagmire and a lake." The reporter continued, "The touchdown today none the less was only one of the burlesque incidents of a game which rapidly became a farce as the rain pelted down and the mud got ankle deep."

The rain began "harmlessly" on Friday afternoon, but became a "veritable cloudburst" on Friday evening. At the weather bureau in downtown Boston, the rain gauge caught 1.48 inches on Friday and 0.50 inch on Saturday — for a total two-day catch of 1.98 inches, a very heavy storm for November. The condition of the field and players was described by a reporter for the *New York Times*:

> The all-night rain left the field in such bad condition that the game was nothing less than an absurd farce. At various spots in the gridiron water had accumulated an inch or more deep. The wonder was that some gladiator was not drowned as he was thrown face down into the water with sixteen or more other youths sitting astride his form. The mud was thick and juicy, and in a few minutes the players were covered with it, their uniforms an indescribable mass of mire, mud in their faces and their ears, and probably in their mouths, mud, mud everywhere.
>
> When a player was tackled he also slid six or eight yards on his back, while the crowd laughed. When the ball landed in the chain of lakes near the centre the water splashed up in the air. The best men were those who had an effective Australian crawl stroke. By winning today Yale clinches the big three football title and took a long stride toward the swimming and water polo titles.
>
> The historians and statisticians say tonight that today was the first time since 1898 that rain had spoiled a Yale-Harvard game. It slackened somewhat in the second period, but still kept coming

down gently, and as the afternoon wore on the mist thickened and it became almost dark, so much so that no mother would ever have recognized her son as being that blackened, mud-spattered miserable looking fellow out in the thick gloom.

One rain in thirty-five years is one too many.

New England's Snow-out: October 10, 1925

New England's earliest significant snowfall in modern records caused a disruption of the Saturday football schedule on October 10, 1925. Snow was reported from all sections of the region: flurries in the extreme south and up to 24 inches in the northern interior. There was considerable drifting in the high winds, and temperatures fell as low as 22°F (-6°C) that night.

Three major college football games were canceled: Holy Cross vs. Providence at Worcester, Massachusetts; Tufts vs. Bates at Medford, Massachusetts; and University of New Hampshire vs. Colby at Waterville, Maine. Many high school games were also postponed. At Hanover, New Hampshire, the game went on: "In the midst of a driving snowstorm which made it almost impossible to distinguish the players, Dartmouth eleven overwhelmed the University of New Hampshire by a score of 50 to 0 today." Down at Cambridge the game went on, too, and apparently stimulated the sons of Harvard, for they rolled up their greatest score since 1891 in beating little Middlebury by 68 to 0. "The icy gridiron was good for the players, although wicked for spectators who sat in seats on which snow had frozen, and in a fifty-mile-an-hour gale," commented a reporter.

Red Grange Comes East — 1925

Red Grange came, saw and conquered. Red Grange came, saw and by his consummate art as a ball carrier shredded the Pennsylvania defense this afternoon and led his Illinois Football team to a 24 to 2 victory over the pride of Pennsylvania . . . He was a wizard, or wild man, or something like that this afternoon, and instead of requiring a dry field upon which to operate, he was fast and elusive and sure-footed on one soaked from rain, deep in mud in places, and dotted with numerous pools of water.

Thus did W. B. Hanna of the *New York Herald-Tribune* describe the famous exhibition of ball carrying displayed at Franklin Field, Philadelphia, on the overcast afternoon of October 31, 1925.

Grange scored three touchdowns in gaining 306 yards and would have made another except that he put the ball down on the one-foot line to permit a teammate to plunge over on the next play and gain a little glory. Twice as he left the field he was given unusual displays of adulation by the 62,000 fans. Near the end of the second half they rose and removed their hats in a silent tribute to his prowess, and again at the end of the game the fans gave him a long, united vocal ovation. For Grange had made football history against watery odds.

Hanna elaborated on his earlier description of the field:

> The gridiron was a mess. When the straw was raked off it before the game, it was indeed an untidy expanse. The rain and snow had soaked in and left it soggy and soft. It didn't appear that men could keep their feet in mire of that sort, but they did and exceedingly well. Evidently such backs as Grange, Daugherty and Briton could run in a swamp. At the end the field was in much better condition than at the start; the trampling of many feet had beaten it down. The Illini, it is said, wanted of all things a dry field, yet they could have cut their devastating capers today on anything from a waxed floor to a lava bed.

Weather conditions certainly were not propitious for such an exhibition of deft footwork. Rain and snow fell on Friday, October 30, to the amount of 0.33 inch. The morning temperature in downtown Philadelphia had dropped to 29°F (−2°C), so that there was little drying overnight with the freezing conditions. No rainfall was recorded on the day of the game, but the temperature rose only to a cool 41°F (5°C).

1925 — Upset at Soldier Field: Northwestern 3, Michigan 2

Coach Fielding Yost of Michigan ranked his teams of 1925 and 1926 as his two best. Using the famed passing combination of Friedman to Oosterbaan, they were accorded top ranking in 1925 along with undefeated Dartmouth. All went well during the season with the exception of a 3–2 upset by lowly Northwestern, played at Soldier Field in Chicago on November 7, 1925, upon a gridiron described as a "morass." Though 40,000 tickets had been sold, only 20,000 fans braved the elements to witness a game that "was played in a sea of mud while a heavy rain was falling."

Both teams fumbled frequently, but Michigan was the chief offender. Both sides kicked many times, Michigan having the edge. A fifty-mile-an-hour wind prevailed, according to a reporter, and "the playing field was covered one inch deep with water." The safety gained by Michigan was a gift from the Northwestern kicker who downed the ball in the end zone on third down as the third quarter ended, in order to secure the benefit of the gale at his back to kick the ball out of danger on the first play of the fourth quarter.

The Chicago weather bureau measured a rainfall of 0.49 inch before the kickoff, and the rain continued throughout the game. The temperature during the game stood at 43°F (6°C).

1934 — The Rose Bowl's Worst: Columbia 7, Stanford 0

Not only was the weather most unusual in Southern California on January 1, 1934, but the result of the annual Rose Bowl game was a shocker also. The Los Angeles area experienced one of its greatest storms on December 31, 1933, and January 1, 1934. The downtown rain gauge caught 7.36 inches of rain within 24 hours, a record that still stands today as the area's most intensive rainfall. Out at Pasadena, where all the festivities take place, the storm total from December 30 to January 1 was 12.86 inches, and much greater amounts fell on the mountains behind the city. The greatest floods of modern times took place on New Year's Eve and Morning. A total of 44 people lost their lives. Communications were disrupted. Only 125,000 people showed up for the annual parade, which had been witnessed by an estimated million the year before. The football game went on as usual, "surging through a drizzling, disheartening rain . . . During the third and fourth periods the rain fell only intermittently. The soggy ball made passing almost impossible and the fact that each team completed a long toss at critical times was astonishing." Columbia with its famous play K-79 scored the only points in the game in the second period, and then held on during the "soggy" second half to win in the greatest upset in Rose Bowl history. They were dubbed the "Sea Lions" by a reporter.

The Twelfth Man in the Snow:
Princeton-Dartmouth — 1935

The entrance of a twelfth man into the lineup of the Dartmouth team when Princeton was threatening to score in an "Ivy League" championship game on November 23 was named by the Associated Press in 1935 as the greatest "sports oddity of the year." It outranked Gene Sarazen's double eagle in the Master Golf Tournament that year. To add to the bizarre incident, a snowstorm was raging, the field was snow covered, and the players were shrouded in the white flakes when the ghostlike twelfth man put in an appearance.

How many men on the field? Dartmouth-Princeton 1935. Courtesy of Princeton University.

The oddity resulted from the fact that the twelfth man was not a player, but a somewhat inebriated spectator who rushed from the stands while the teams were lining up on the three-yard line during the snowstorm. Out of the swirl of flakes came the supernumerary. Though seen by the umpire and the safetyman, neither was able to prevent his plugging a hole in the Dartmouth line that the Princeton back was about to plunge through to the goal line.

The snow scene was vividly described by Frank D. Halsey in *The Princeton Alumni Weekly*:

> It might be explained that the contest took place in a near-blizzard. Snow had begun to fall early in the morning, and fell heavily from before the kickoff until after the final whistle. Although the field had been protected by tarpaulins until the last possible moment, it was dusted white before the first quarter was over, and was an inch deep with snow before the start of the second half. The temperature was just low enough to keep the snow from melting, yet high enough to keep it moist and make the playing surface treacherous and slimy — much worse than any honest rain would have made it. There was also a stiff breeze from the north which made it uncomfortable for spectators and players alike, but especially for the team defending the south goal.

The snowstorm was caused by a low-pressure center well off the coast of New Jersey. Though traces of snow fell as far south as Philadelphia and northward over New York City, only in central New Jersey did much accumulate, and Princeton was apparently in the center of the snowbelt. Trenton reported 0.21 inch and New Brunswick 0.19 inch of precipitation, which would have produced about 2.0 inches of snow. Minimum temperatures at both places that day were 29°F (-2°C), and the thermometer remained just below the freezing point during the game.

After the twelfth man was escorted from the field by the police, Princeton proceeded to make its touchdown and to end the game a winner by 26–6 over previously undefeated Dartmouth. The Tigers threw only one pass all day while grinding out four drives of 42, 48, 40, and 50 yards for scores over the snow-covered terrain of Palmer Stadium.

The identity of the twelfth man drew much attention in the

press because there were two claimants: Mike Mesco, a short-order cook from Rahway, and George Larsen, a young architect from Cranford. An authority on Princeton football, Jay Dunn in *The Tigers of Princeton*, has concluded: "But no one has ever proven conclusively whether the 12th man was Mesco or Larsen. It might have been either. Perhaps it was neither."

Football's Worst Saturday Weather — 1950

The fourth Saturday in November often is the big day in intercollegiate football when conference and league championships are decided in a big game. In 1950 this Saturday was two days after Thanksgiving Day, so many fans combined the annual Turkey Day festival with a Saturday football game for a big weekend. These same days produced the most spectacular winter-type storm to hit a wide section of the country east of the Mississippi River during this century. The Great Appalachian Storm of 1950 appears to stand unique in the recorded weather annals of the region for the combination of severe weather elements, all raging at their extreme fury.

A small storm center developed over western North Carolina and Virginia on Friday morning, November 24, and after a southward loop over North Carolina headed northward with its rapidly deepening center tracking just to the east of the crest of the Appalachian Mountains on Friday night and Saturday morning. At kickoff time, 1:00 P.M. (EST), the center was near Harrisburg, Pennsylvania, and about to make a northwestward turn that would take it into northern Ohio soon after midnight.

For sections west of the storm track and the mountains, this atmospheric whirl caused the coldest weather for so early in the season ever known and dropped the deepest November snows ever measured. East of the mountains, in the warm sector of the cyclonic storm, wind and rain prevailed. The highest wind gusts ever experienced in a nonhurricane situation were reported at a number of stations, and some of the heaviest November rains of record fell over Pennsylvania and New York.

As it happened, that was the day when the Western Conference title was to be decided at Columbus, Ohio, at the game between Michigan and Ohio State, while at Princeton, New Jersey, Dartmouth and the Tigers were tangling for the mythical Ivy League title. Each had its own variety of weather extremes.

Heavy snow fell throughout the game at Columbus. "So heavy was the snow that at times it was impossible to see the players from the press box, and when measurements for first downs were necessary, brooms were used to sweep off the grid-iron and find the yard markers," reported the *New York Times*. Despite the atrocious conditions, only one fumble was made. Forty-five punts turned it into a kicking game. Michigan scored its points as a result of blocked kicks, winning by 9–3. Although 82,300 tickets had been sold, only 50,503 braved the elements to see the game. The Columbus weather bureau measured 7.5 inches during the day, and a total of 9.0 inches was on the ground in early evening. The thermometer tumbled throughout the game to reach a low of 10°F (−12°C) that night.

At Princeton, it was a different story. The game was played "under weather conditions so miserable as almost to defy description," wrote Joseph H. Sheehan of the *New York Times*.

> A howling gale out of the east and lashing rains turned the field into a quagmire, and kept away from Palmer Stadium all but 5,000 hardy enthusiasts of the 31,000 who had purchased tickets and made this a travesty of a football game. However, the brilliance of Charley Caldwell's mighty single-wing machine shone through the murk and mire to convince all on hand of a superiority over the Indians that undoubtedly would have been greater under more favorable circumstances.

With All-American Dick Kazmeier scoring two touchdowns, Princeton won the game by 13–7 and the mythical crown of the yet-to-be formulated Ivy League.

The cold front driving from the southwest reached Princeton about 3:30 P.M., early in the fourth quarter. The gale-force winds quickly lost their energy and the torrential downpours slackened off considerably. While still in the warm sector, the Newark Airport measured a gust of 108 mi/h (174 km/h) and there were other stations in the Northeast that topped 100 mi/h (161 km/h). Temperatures in central New Jersey ranged up to 61°F (16°C) in the warm sector, and that was the approximate temperature during the first half of the game at Princeton. Local records from surrounding communities show that from 2.13 to 2.75 inches of rain fell during the morning and early afternoon of November 25.

Nearby Rutgers canceled its game with Colgate scheduled for that afternoon at New Brunswick. The University of Pittsburgh postponed its big game with Penn State. Twelve inches of snow fell at the Pitt Stadium on the twenty-fourth and 10 inches more on the twenty-fifth. There were 20 inches on the ground at 7:00 P.M. on Saturday evening.

National Football League Championship Games

The season of professional football extends from the sweaty exhibition games of midsummer to the often congealing climate of the final playoffs in late December and early January. All sorts of weather can happen in this extended period, and usually does on a Sunday afternoon.

Commencing in 1933, the championship of the National Football League was played in mid-December at the home field of the team with the best record. The second game of the series, played on December 13, 1934, achieved some notoriety and was remembered as the "Sneaker Game," not from any hidden ball play, but as a result of the substitution of rubber soled shoes for regular cleats by the Giants' coach. The day before the game, the temperature in New York City was down to 10°F (−12°C), causing the playing surface to freeze solidly. The use of sneakers in the second half enabled the host Giants to run and cut on the icy surface and outplay the visiting Chicago Bears, who had brought along only shoes with leather cleats.

The championship game at Cleveland on December 16, 1945, was played in as cold an atmosphere as one can get at that locality at this early date in prewinter. The temperature ranged from −3°F (−19°C) in the early morning to an afternoon maximum of only 8°F (−13°C). "It was so cold that the musical instruments froze," claimed a reporter. The Cleveland team proved to be the better polar bears by edging the Washington Redskins by 15–14. It might be mentioned that this was the same date in

December 1948: Philadelphia Eagles 7–Chicago Cardinals 0. National Football League Properties, Inc.

1835 that brought "Cold Wednesday," with the bitterest daylight temperatures and coldest wind chill ever experienced in the Northeast.

The championship game at Philadelphia's Shibe Park on December 19, 1948, was played under "almost impossible blizzard conditions," and was ever after dubbed the "Blizzard Bowl." A heavy snowfall began in the morning and when the workmen removed the tarpaulin from the field all the chalk lines were frozen fast to the cover and rolled up with it. The kickoff was delayed an hour in order to plow the field and stack the snow away from the playing field. A rope was stretched along the sidelines to mark the borders of the field. During the early part of the game, snow fell at the rate of 2 inches an hour. The weather bureau's official measurement for the day was 7 inches. The temperature ranged from 27° to 31°F (-3° to -0.6°C). Despite all the handicaps the game was remarkably well played. For three frigid periods the 28,864 spectators who braved the weather watched the teams battle back and forth "as though competing in some Eskimo frolic." Finally, after a fumble Steve Van Buren mushed over for the only score of the game, giving the Eagles a 7–0 victory over the Chicago Cardinals.

The wettest championship game was played at Los Angeles on December 19, 1949, when the day's rainfall catch measured 1.86 inches. During the hour from 1:00 to 2:00 P.M., a whopping

0.22 inch fell. The Eagles outswam the Rams by 14–0, showing that they could operate on either snow or water. Only 22,245 fans found their way to the 100,000-capacity Memorial Stadium. The maximum temperature in Los Angeles was 58°F (14°C).

Green Bay in northeast Wisconsin at latitude 44°30' North used to be called the "North Pole" of the National Football League, mainly as a result of the playoff and championship games that were played there in the 1960s when the Packers were a powerhouse. "In a roofless igloo, known as City Stadium," Green Bay demolished the New York Giants on December 31, 1961, by 37 to 0, with the temperature that day ranging from 6° to 23°F (−14° to −5°C) and the sidelines covered by 7 inches of snow. "The turf, kept covered this morning, provided good traction for straight ahead running, but its frozen surface made it difficult for runners to cut," observed the reporter of the *New York Times*. Allie Sherman, the Giants' coach, said: "No, the weather wasn't a factor." The slipping and sliding was about even, he thought, but when a team gets ahead in such a situation they are much more relaxed and need not press on a tough playing surface.

At the game the next year, at Yankee Stadium in New York City, the Green Bay Packers not only brought themselves, but they also brought their climate along. The spectators on December 30, 1962, "came bundled in parkas and greatcoats and blankets and most of the 64,892 of them yelled 'Beat Green Bay' through the gelid afternoon. But in the end, making their way through the early dark and swirling wind to the subway, they accepted the sad truth: this was not theirs," philosophized the *New York Times* reporter. The temperature that day took a tumble from an early 35°F (2°C) to an evening 5°F (−15°C); at game time it read 18°F (−8°C) and dropped to 10°F (−12°C) by the final whistle. The wind gusted as high as 40 mi/h (64 km/h) at 3:00 P.M., and blew at a steady rate of 26 mi/h (42 km/h). Though 0.35 inch of rain had fallen on the twenty-ninth with a trace of snow, none came on the day of the game and there was no snow on the ground as the arctic blasts swept the bare playing field.

Next year, 1963, conditions were only slightly better at Chicago, when the Chicago Bears met the New York Giants on December 31. Temperatures at a weather station near the stadium ranged from a morning 3°F (−16°C) to an afternoon 20°F (−7°C). At O'Hare Airport, just north of the city, it had dropped to −14°F (−26°C) overnight. There was only a trace of snow on the sidelines. The field was frozen but playable. Chicago won by 14–10.

Green Bay: 1965 NFL Championship

"Playing through their hometown snow, sleet and slush like a herd of happy polar bears, the strangely wonderful Green Bay Packers [won] on a field less conducive to the sweep and cut back because it was soft and slippery," wrote Tex Maule in *Sports Illustrated*. This was the championship of the 1965 season, which Green Bay won by defeating Cleveland by 23–12. The game was played on January 2, 1966.

Arthur Daley in the *New York Times* gave a more detailed account:

> Three inches of snow fell in the morning, but the field was clear when the tarpaulins came off shortly before game time. Then the snow moderated and changed to sleet as the field became slick. But the players did not consider themselves to be badly handicapped . . . At least it was not the worst of these snow-festooned classics,

he concluded.

Snow fell throughout the morning; the measurement at the Green Bay Airport was 4.1 inches. At game time the temperature was just above freezing at 33°F (1°C), having risen from an overnight low of 20°F (−7°C).

Green Bay: 1967 NFL Championship

> If ever there was a game played under more execrable conditions than the Packer-Cowboy showdown, it was too insignificant for posterity to remember. This had to be the worst ever. Even the "electric blanket," the 14 miles of wiring that heated the gridiron, was powerless to combat the penetrating cold. The blanket did prevent the field from becoming a skating rink, but the gridiron grew ever more slippery as the day wore on until cleats stopped biting and traction was lost,

reported Arthur Daley in the *New York Times*.

The Green Bay Packers, with a record of 9 and 4, managed

to win the league semifinal playoff against Baltimore and then returned to their northern homegrounds to clash with the Dallas Cowboys for the National Football League championship on December 31, 1967.

A severe cold wave was descending on the upper Great Lakes region that weekend, with a strong anticyclone bringing a frigid mass of Canadian air southward on the thirtieth and thirty-first. In the advective blasts of the northwest winds, the mercury at Green Bay stood at 6°F (−14°C) at midnight of the thirtieth, and continued dropping throughout the following day to reach −19°F (−28°C) at the time celebrants were ushering in the New Year. At kickoff time the Green Bay thermometer stood at −13°F (−25°C). A number of stations in Wisconsin reported the mercury at −30°F (−34°C) on New Year's morning.

Fortunately, it was a dry period. At Green Bay Airport, 0.01 inch of snow fell on the twenty-eighth and 0.03 inch on the twenty-ninth, both insignificant amounts. No precipitation was recorded on the thirtieth and thirty-first. Though there was a snow cover of 2.0 inches at the airport on game day, the playing field had been thoroughly cleared.

Arthur Daley concluded his column on the spectacular doings:

> Thus did the Packers win, 21–17, and whisk themselves from the frozen tundra of Green Bay to their destined date with the Oakland Raiders in the Super Bowl at more salubrious Miami a fortnight hence. It was an unbelievable finish to an unbelievable game. Not since Princeton and Rutgers invented the sport almost 100 years ago has there been anything to compare with this one. Historians will babble about it for decades as one of the fantastic episodes in football annals.

Super Bowls

Upon the merger of the National Football League and the American Football League in 1965, the Super Bowl came into being and found a peripatetic home in the Sun Belt, far from the inclement weather that is experienced at this season in the home cities of the northern teams. Only the three games played outdoors at New Orleans in the Tulane Stadium have been marred by adverse weather or threatened by potentially adverse conditions. The only other city to host as many Super Bowl games was Miami, where January conditions are generally favorable for outdoor activities. Three games were played in Southern California under good weather: two in the Coliseum in Los Angeles and one in the Rose Bowl at Pasadena. Super Bowl VIII was played at Houston under fair skies.

Super Bowl IV, played at New Orleans between the Kansas City Chiefs and Minnesota Vikings on January 11, 1970, was preceded by a night and morning of heavy rain and thunderstorms. Tornado warnings were issued for the afternoon of the game. Fortunately, the cold front moved through about noon, the rain stopped, and the temperature mounted to 64°F (18°C) about game time. The field was described as "a combination of mud, sand, and green dye" that was most spectacular on color television. The rainfall at the downtown weather station measured 1.34 inches. The Chiefs won by 23–7.

Super Bowl VI at New Orleans on January 6, 1972, should have been dubbed the "Cold Bowl Dixie-style." The temperature at kickoff stood at 39°F (4°C), up from a chilling overnight 24°F (–4°C). There was bright sunshine, however. The Dallas Cowboys defeated the Miami Dolphins by 24–3.

Super Bowl IX on January 12, 1975, also played at New Orleans, was preceded by 0.81 inch of rain on the tenth and 0.12 inch fell on game day. The Pittsburgh Steelers arrived with special rubber-cleated shoes, which apparently helped them to defeat the Minnesota Vikings, 16–6.

The Los Angeles Coliseum was the scene of the two warmest outdoor Super Bowl games. At the first, on January 15, 1967, the temperature reached 79°F (26°C) on the downtown weather station thermometer. This was topped at Super Bowl VII on January 14, 1973, when the peak of 82°F (28°C) was attained; this was 16°F (9°C) above the normal maximum for the date.

Now, with the Super Bowl being played indoors under domes in most years, the prevailing weather outside is of little importance in determining the quality of the play.

The first National Lawn-Tennis Tournament, on Staten Island. Courtesy of the Library of Congress.

Tennis

Origins of American Tennis

Tennis is thought to be of Oriental origin, later adopted by Greeks and Romans, reintroduced into France by the Saracen invasions, and developed in the late Middle Ages into court tennis, which bore a vague resemblance to the game we play today. The form of lawn tennis with standard court measurements and established rules was the work of British Major Walter C. Wingfield, whose comprehensive book, *The Major's Game*, was published in 1874.

The question of who was the first to lay out a tennis court in the United States has been the subject of much argument and recent research, but most authorities now concede precedence to Miss Mary Outerbridge, who returned in the late winter of 1874 from Bermuda to her Staten Island home with a set of racquets and a net, along with a rule book. She obtained these from British officers who had recently brought the game to Bermuda. Soon after her return a court was laid out on the grounds of the Staten Island Cricket and Baseball Club at Camp Washington, now St. George and part of New York City. Gentlemen and ladies were soon at play.

Another court, of greater historical significance, was laid out at Nahant, a resort northeast of Boston, most probably in the late summer of 1875, and there James Dwight, the "Father of American Tennis," played his first game. His opponent was Fred Sears, the older brother of a coming perennial American champion. Dwight described the game, "I remember even now that each won a game, and as it rained in the afternoon, we played in rubber boots and coats rather than lose a day."

Harper's Weekly, in an 1878 article describing the new game and its rules, had the following to say: "Balls covered with white cloth should be used in fine weather. In wet weather a hollow India Rubber ball without any puncture or hole is preferable." Apparently, the game was played rain or shine.

The first tournament in the United States took place at Nahant in the summer of 1876, when fifteen players participated in a round-robin affair that was won by James Dwight.

National Tennis Championships

The men's United States National Tennis championships were staged at the West Side Tennis Club in Forest Hills, New York City, from 1915 through 1977. Of late they have been scheduled to cover twelve days during the first half of September. The play was open only to amateurs until the 1968 tournament, when the professionals joined to make the United States Open Championship.

During the number of days required for the tournament, several changes in the type of weather are to be expected. It is normally a dry period, since cyclonic activity is at a seasonal minimum, but there can be rain from three other sources: from a tropical storm moving northeastward along the seaboard, from a stalled frontal system that refuses to move along, and from an occasional afternoon thunderstorm.

The heaviest twenty-four-hour rain ever to fall during the tennis period was 4.86 inches, when the "Morro Castle" storm from the tropics struck on September 8, 1934. The storm derived its name from the passenger liner that burned at sea and was stranded at Asbury Park, New Jersey, during the storm.

Temperatures as high as 102°F (39°C) have occurred at this time of year, according to New York City records. This was in 1953, fortunately before the matches had begun. Only once before had the mercury exceeded 99°F (37°C) in the first week of September; this came 'way back in 1881, on the day after the famous Yellow Day. Over the years, readings of 90°F (32°C) or more have occurred on each of the first fifteen days of September, but the normal daily maximum at the middle point of this period, September 7, is 79°F (26°C).

For data prior to 1948 we turn to the Weather Bureau station at Battery Park in lower Manhattan, located about 8.5 miles west-southwest of Forest Hills. From 1948 to the present, the data are from La Guardia Airport, 4 miles north-northeast of Forest Hills. Beginning in 1978, the matches have been staged at the new stadium in Flushing Meadows, only 1.5 miles from the weather station at La Guardia Airport.

1934

In addition to the heaviest one-day rain, the worst string of bad-weather days to plague a championship tournament since it was moved to Forest Hills in 1915 hit the West Side Tennis Club Stadium in early September 1934, when there were three full-day postponements, three days with curtailed schedules, and one when play was interrupted twice by violent local storms. This was the time of the Morro Castle disaster, mentioned previously.

The tournament opened on September 1, then rain caused postponements on the third and fourth days, when 0.46 and 0.37 inch fell. Play was resumed on the sixth, but intermittent raindrops caused a curtailment of the program. On that day and the next, 0.13 and 0.26 inch dropped from the leaden sky. The eighth was a complete washout, as this was the day the tropical storm crossed the central Long Island coast and Forest Hills lay in the zone of maximum rainfall.

Next day, toward the conclusion of a close match between Sidney Wood and Frank Parker, a violent thunderstorm broke over the Stadium "to create one of the worst scenes on record at a national tournament," according to Allison Danzig, the tennis dean of the *New York Times*. A crowd of several thousand people rushing for the exits at once created a terrific human traffic jam. No sooner had they returned to their seats and play resumed than a hailstorm broke in a "sun shower" as the beams of the setting sun were still gleaming on the hailstones. After twelve scheduled days the tournament finally ended three full days behind schedule. Fred Perry of England defeated Wilmer Allison of Texas in the final.

1935

The tournament in 1935 started off with a rain-out on the first scheduled day, August 30, when only 0.08 inch was measured. Saturday and Sunday saw sunny skies and a full schedule of play, but Monday brought rain and a curtailed program. This was the preliminary to Hurricane No. 2 of the 1935 season, which dumped 3.60 inches during the four days from Monday through Thursday, September 2 to 5. The wet conditions of the court on Friday caused another full postponement of play. So five vital days were lost. The weekend went on schedule, but rain again on Monday, the ninth, curtailed play. Not until Wednesday, the twelfth, after a record span of fourteen scheduled playing days, did the tournament conclude, with Wilmer Allison of Texas defeating Sidney Wood of New York.

1969

The 1969 tournament opened early on Wednesday, August 27, and proceeded smoothly for the first week. Rain fell on September 2 during a match between John Newcombe and Marty Riessen, causing a delay of one hour and ten minutes in play. This resulted in a postponement of the Pancho Gonzales and Tony Roche match, for which the crowd had been eagerly waiting. Rain during the day measured 0.22 inch, but this was only a prelude to the downpours that followed on the third and fourth (2.33 and 3.12 inches) washing out the entire slate on those days. A stationary front had stalled over the New York area. Though more rain fell on the sixth (0.25 inch) and on the eighth (0.18 inch), the tournament concluded on Monday, the ninth, only one day late. Rod Laver won the national championship for the second time by defeating Tony Roche.

1971

The tournament of 1971 took longer to complete than any other. Play progressed smoothly until the eleventh day, September 11, when it was washed out by a morning rain of 0.48 inch; though the rain stopped before 2:00 P.M., play was not begun. On the twelfth, the rain increased to a fall of 1.31 inches, on the thirteenth to 0.70 inch, and on the fourteenth to the even heavier amount of 1.94 inches, for a total of 4.43 inches during the four days. A slow-moving frontal system caused the excessive downpours. The semifinals on the fourteenth were played in a misty rain. The final came on Wednesday, the fifteenth, three days late. Stan Smith beat Jan Kodes. Much restiveness was displayed during the final days by many players, especially those required to stay for the doubles. Most professionals had prior remunerative commitments for the week in other cities.

1974

The 1974 tournament also ran into some rain difficulties. The second and third days had curtailed programs, but then a solid six days of completed play brought the matches along. Despite heavy rains on September 1, 2, and 3, for a total of 2.81 inches, all matches were played. Then, "a steady downpour" on September 6 and 7, dropping 1.28 inches, caused postponements on Friday and Saturday. This put the final over to Monday, the thirteenth scheduled day, when Jimmy Connors beat Ken Rosewall for the title.

National Women's Tennis Championship

A weather incident may have played a decisive role in the outcome of a key match in the National Women's Championship in September 1950. Althea Gibson, the first black to win national fame in tennis, was within four points of victory in a hard-fought contest with Louise Brough when a thunderstorm broke over the stadium. "The drenched players ran for the locker rooms. The fans scurried to shelter under the stands. A bolt of lightning toppled one of the great concrete eagles above the stadium but luckily no one was beneath as it crashed to the ground," reported the *New York Times*. The conclusion of the match was postponed until the next day when Louise Brough regained mastery and won by the scores of 6–1, 3–6, and 9–7. "The elements robbed her of her great triumph," the reporter commented on Althea Gibson's near victory.

Davis Cup

The worst weather experience in all Davis Cup history on United States soil occurred in the 1903 matches played at the Longwood Cricket Club in Chestnut Hill, Massachusetts, and the intervening rainfall had an important part in deciding the challenge round. This time the British sent their best team, consisting of the Doherty brothers, Reggie and Hugh. Reggie injured his arm in an early practice and was not permitted by his physician to play in the opening matches scheduled for Tuesday, August 4. Hugh won the first singles, but Reggie's match was defaulted. A heavy rain of 1.61 inches fell overnight and continued on Wednesday, causing a postponement. When rain fell again on Thursday, the doubles match was put over to Friday. With three days extra rest, Reggie returned to the court with his arm healed and helped to win the doubles. Each brother won again on Saturday in the singles to take the cup back to England where it remained until its first "Down Under" trip in 1907.

Golf

National Open Golf Championships

> In most of the spectacular championships in American golf history
> it had rained or been disagreeable weather.
> — Harry B. Martin, *Fifty Years of American Golf (1936)*

Though golf made an early start in America, it did not take hold
as a popular sport until a relatively late period. The game appears
to have been played in the South in the years immediately following the end of the War of the Revolution. The Harleston Green
Club was organized at Charleston, South Carolina, in 1786, and
notices of meetings of a golf club appeared in Savannah newspapers at this time. The organizers were of Scottish descent and
imported the game from their native land, where it had flourished
for years. Golfing activity seems to have continued in this region
for about twenty-five years and then all mention of it ceases.

Not until the 1880s did the game surface again within the
United States, though it had been played in Canada, where clubs
were organized in the previous decade. Golf courses were laid out
in such rural retreats as West Virginia and Kentucky in the middle 1880s. Game activity in the New York City area dated from
February 1888 when John Reid, the "father of American golf,"
invited some friends to play on a course he laid out in a cow pasture on Broadway in Yonkers, just over the city line of New York
City. As a result, a meeting was held in November of that year
and a golf club organized, appropriately named the St. Andrews
Club. In the early winter of 1894, members of this group founded
the United States Golf Association, which has guided the sport
over the years to the eminence it enjoys today.

The first semblance of a national tournament took place in
October 1895 at Newport, Rhode Island, when both amateurs and
professionals staged contests on successive days. H. B. Martin in
Fifty Years of American Golf has related that "when the matches
got under way on a cool October morning there was a sharp
northwest wind blowing which considerably reduced the size of

the gallery." Horace Rawlins, a 19-year-old, with 91-82-173 won on his home course and took the first-prize money of $150, in a total purse of $350. Four rounds (36 holes) were played in one day on the nine-hole course.

The second Open Championship of the Golf Association took place on July 16, 1896, at the Shinnecock Hills Golf Club at Southampton on Long Island. A New York reporter was on the scene to cover the event. He reported that a thunderstorm set in about daybreak and had the beneficial effect of clearing the atmosphere. "About a half hour before the time set for the beginning of play, the weather began to clear, and at 10:10 it was beautifully clear." Later he reported: "The weather is perfect for golf, a delightful breeze blowing from the west, but not strong enough to interfere with the drive. The sky is azure, the dark blue of the water, and the rich green of the turf set off the red coats and bright plaids of the golfers, and with the beautiful dresses of the women, make a brilliant picture." James Foulis of the Chicago Golf Club was the winner by three strokes over Horace Rawlins.

Varied Climates of the Open

The 83 National Open Golf Championships played through 1983 have been staged in a variety of climate zones: from Cape Ann in Massachusetts on the Atlantic Ocean to Pebble Beach in California on the Pacific Ocean; from Minneapolis in the North to Houston and Atlanta in the South; and from mile-high Denver to sea level or thereabouts on both coasts. The far-flung championship contests point up the varieties of our American climate.

Over the more than eighty years of the National Open Championships, two tournaments stand out for the effect of weather on the contestants and the playing scene. The spectacular victory of Francis Ouimet at The Country Club in Brookline, Massachusetts, in 1913, has long been enshrined in the annals of golf as the greatest upset in the national series, and the greatest comeback in golf history resulted from the dogged endurance of Ken Venturi at Bethesda, Maryland, in 1964, when, though almost overcome by the heat and humidity during the morning round, he returned to struggle through a winning afternoon round in which he gained six strokes on the leader and won.

1913

The most dramatic finish in Open history occurred in an exciting playoff, when Francis Ouimet, a 20-year-old Massachusetts amateur, defeated Harry Vardon and Edward Ray, two of the most famous names in British and international golfing. Ouimet had persevered through a rainy afternoon on the last round to sink a putt on the eighteenth hole and win a tie with Vardon and Ray, who were already in the clubhouse. In the playoff, also under wet conditions, Ouimet triumphed by steady play with 72 against Vardon's 77 and Ray's 78. Ouimet became only the second American-bred golfer to win the Open Championship.

The Boston area was enduring one of its September rainy spells that sometimes mar the otherwise beautiful autumn season. These are caused by prolonged northeasters precipitating Atlantic Ocean moisture over the land. The matches ran from September 16 to 20, with rain falling on each day of play except the first. A reporter from the *Boston Evening Transcript* kept his eye on the weather:

> Sept. 16 — "Conditions were ideal."
>   " 17 — "A few sprinkles early in the day."
>   " 18 — "Rain in the morning kept the crowds down to slim proportions."
>   " 19 — "Worse conditions than on any previous day prevailed this morning, the early starters in particular getting caught in heavy showers. A fresh rainstorm on top of that of the night before last still further softened the turf."
>   " 20 — "Heavy mist and rain did not seem to make any difference to thousands of Golf enthusiasts."

A reporter for *American Golf* magazine was on hand to make some further observations of playing conditions on the last two days:

> Friday [the nineteenth] dawned, or rather didn't dawn, because a worse day for golf has rarely been seen. The rain was cold, and the links all covered with heavy moisture, presenting anything but an attractive sight to Ray and Brady as they stepped to the tee.

The day [the twentieth] was bad for golf and bad for spectators. It was raining: had been raining for two days ... The ground was soggy, putting a premium on length and accurate mashie work and making putting rather difficult.

Writing in the same magazine, Bernard Darwin, a British commentator, observed: "No golf could have had more cruel and pitiless weather to withstand, but [the course at] Brookline came out of its ordeal with flying colors."

A weather station located at Chestnut Hill, near Brookline, recorded the following maximum temperatures and precipitation on the days of play:

Sept. 16 — 65°F (18°C), no rain.
   "   17 — 74°F (23°C), 0.55 inch.
   "   18 — 77°F (25°C), 0.10 inch.
   "   19 — 58°F (14°C), 0.55 inch.
   "   20 — 59°F (15°C), 0.18 inch.

1964

The nation's capital on the low-lying Potomac River has long been known for its hot and humid summer weather, and it provided a classic example of this at the second National Open staged in its vicinity. The first day, June 18, was about normal, with a maximum of 84°F (29°C), on the second day the thermometer rose to 90°F (32°C), and on the final round it hit a sizzling 95°F (35°C). These readings were on the official thermometer of the Weather Bureau at the Soldiers Home in nearby Rockville. Lying in the warm sector of a cyclonic system whose center was over the lower Great Lakes, the Washington area was under the influence of hot, humid airstreams from the Gulf of Mexico. Both the relative and absolute humidities were exceptionally high.

The final "brutally hot and humid Saturday" provided the scene for a dramatic comeback by Ken Venturi, who had almost faded from golfing notice after a spectacular year in 1960. Six strokes behind after 36 holes on Friday, Venturi turned in a dazzling 30, five under par, for the first nine on Saturday morning, and was six under par on the seventeenth and eighteenth holes. But the heat had taken its toll. He appeared exhausted after his morning round of 66, and there was doubt as to whether he would be able to face more heat and humidity.

A physician member of the Congressional Club, Dr. John Everett, advised only tea and salt for the noon intermission to counteract his dehydration. Venturi did so and obtained the unusual permission of having Everett accompany him on the afternoon round. The doctor carried a plastic bag full of ice for the golfer's use.

"Slow it was, and slower it got. Dripping wet from head to shoe, the 33-year-old Venturi trudged from hole to hole, from shot to shot, often stopping altogether, then barely pushing one foot in front of another. But by some miracle his golf was not affected," commented the *Washington Post* on the afternoon round. Venturi managed to par the final four holes for a card of 70 and finished four strokes ahead of Jacobs, who had wilted under the heat to a 76 on the final round. The *Post* commented again: "A crowd of 17,104 was barbecued in 92-degree temperature and 85 percent humidity."

Coolest

The coolest playing day in National Open history occurred in a memorable championship at the Brae Burn Country Club in West Newton, Massachusetts, on June 9, 1919. The day was rainy with 0.32 inch falling at nearby Chestnut Hill, and the temperature ranged from a low of 50°F (10°C) to a high of only 52°F (11°C). It warmed up next day to 62°F (17°C) and to 70°F (21°C) on the final two days. Another chilly playing day occurred at The Country Club in Brookline, Massachusetts, on September 19 and 20, 1913, the famous victory of Francis Ouimet. The last day of the tournament and the playoff day were subject to wet conditions, with afternoon temperatures of 58°F and 59°F (14°C and 15°C).

The 1902 Open at Garden City, New York, on Long Island, experienced a cool day on October 10 when the maximum temperature reached only 57°F (14°C).

The only other place in the country to produce playing temperatures under 60°F (16°C) was the Olympic Country Club in San Francisco, for the contest in 1955, when readings of 58°F and 59°F (14°C and 15°C) prevailed on the last two days.

The annual Bing Crosby tournament in January 1962 was played at Pebble Beach, California, in ice, snow, sleet, hail, and rain. Monterey Peninsula Herald Co.

Hottest

The record for the hottest conditions must go to the Southwest, where 90°F and above are the normal in June. Dallas appears to hold the record, with 98°F (37°C) on two days in June 1952, and these were preceded by days with readings of 95°F and 97°F (35°C and 36°C). Minneapolis in 1930 hit 98°F (37°C) one day, but cooled on the following days. Glenview, Illinois, during the famous heat wave of 1933, had two days with 97°F (36°C). Toledo in 1931 had a string of four days with 90°F (32°C) or more.

The most uncomfortable atmospheric conditions, however, usually are met on the East Coast when a Bermuda High produces very humid weather along with high temperatures. At Englewood, New Jersey, in 1909 the mercury topped at 94°F and 95°F (34°C and 35°C) along with very high humidity. The final two days at the Congressional Country Club at Bethesda, Maryland, in 1964, endured heat of 90°F and 95°F (32°C and 35°C) accompanied by a "Turkish bath atmosphere." The 1925 Open at Worcester, Massachusetts, was played during a memorable early June heat wave when the thermometer rose to 92°F (33°C) on each of the three playing days.

When asked about the heat at Englewood in 1909, Tom Vardon, a brother of the British champion, said: "After I had played about six holes in the morning, I felt certain I would never finish. I've never experienced such heat before, and I don't want to again."

"Why didn't you take off your coat?" a reporter asked.

"Take off my coat! I wouldn't think of doing such a thing on the other side."

Next day, however, the reporter had this to say about Vardon: "He dropped his coat yesterday and charmed everyone by a pea green shirt, lighter in tone than his green felt Alpine hat."

Wettest

The eighty-third Open Championship, played at Oakmont near Pittsburgh, suffered the worst rain disruption in the history of the contest. On Friday, June 17, 1983, the second-round play was suspended by darkness at 8:40 P.M., with 38 of 155 players still on the course. An afternoon thunderstorm had caused a delay of 2 hours and 32 minutes. The round was finished on Saturday morning.

For the first time in history, the final round on Sunday was not completed that day. Rain began to fall and lightning threaten, so play was halted at 5:29 P.M. with only 6 of the 70 who started the round still on the course. The rain continued, and a national television audience had to wait until Monday to see the conclusion.

Lightning

The National Weather Service has made an attempt to minimize the dangers to professional golfers due to lightning. At the Public Golf Association tournament at Tulsa, Oklahoma, in August 1982, PGA officials through the use of storm spotters and radar kept track of any potential thunderstorm. Their plan is to display a green flag when informed the weather was good for play, a yellow flag if a storm with lightning was possible within two hours, and a red flag if lightning was likely within 30 minutes. With a red flag flying or lightning actually visible, players can stop play and head for shelter on their own volition.

The Kentucky Derby

"The most exciting two minutes in sports" was first run at Churchill Downs in South Louisville on May 17, 1875. Since then the premier horse racing event of America has been held on dates varying from April 29 to May 23, with the exception of the war year of 1945, when the date was June 9. In recent years, the annual spectacle has always been staged on the first Saturday in May.

Louisville weather records for the past century place the normal maximum temperature during the first week of May at either 72° or 73°F (22° or 23°C). For the past 46 years, when the race has been on the first Saturday in May, the odds are even that the temperature will be between 65°F (18°C) and 81°F (27°C). During this period the thermometer failed to reach 65°F on only nine occasions. Derby Days in 1937 and 1957 hold the dubious honor for the coolest conditions with an afternoon maximum of a chilly 47°F (8°C). The hottest day was May 2, 1959, with a sizzling 94°F (34°C).

The statistics show that there is a 40 percent chance of rain during the 24 hours of any day during the first week of May, but the afternoon hours between 1:00 and 7:00 P.M. have only a 25 percent chance of rain in some form. The wettest Derby Days were in 1918 and 1929. An inch fell during the afternoon of May 11, 1918, and 2.31 inches during the 24 hours. The rainfall on May 18, 1929, amounted to 1.19 inches.

Of the 109 races staged through 1983, 80 percent have been run under favorable conditions, with the track classified as clear, fast, or good, while 20 percent have had the track judged heavy, muddy, slow, or sloppy. Several of the poor track conditions have been the result of rain on previous days and not on Derby Day.

During the first 25 years of the classic, there were only four years when the track was not fast or good. Since the turn of the century, poor track conditions have come in bunches: 1905–09, 1918–20, 1925, 1927–29, and 1945–48. Since 1948, a span of 35 years, the track has been either fast or good in all years with the single exception of 1958, a remarkable record for a region where frontal passages are frequent and thunderstorms possible at this time of the year.

May 10, 1905
"The rains of last night and the terrific downpour of this morning have made the course heavy and slippery." — *Louisville Courier-Journal*. In this heavy track, Agile won easily in 2:10 3/4.

Rainfall amounted to 0.03 inch on the ninth and 0.58 inch on Derby Day. The maximum temperature rose to 89°F (32°C) in the afternoon.

May 6, 1907
"The race was run over a track almost fetlock deep in mud, and the time, 2:12 3/5, was the slowest in the history of the race." — *New York Herald*. The winner, Pink Star, trailed far back till the three-quarter pole, and won at odds of 15 to 1.

Rain fell to a depth of 0.26 inch on the fifth and to 0.67 inch on the sixth. It was falling at 7:00 A.M. on the day of the race. The maximum temperature that afternoon was 67°F (19°C).

May 3, 1909
A fast race despite the slow track, with Wintergreen taking an early lead to win in 2:08 1/5.

"Early in the morning a heavy rain began to fall and this continued to almost noon, keeping away from the track many thousands. More than ten thousand saw the contest in spite of the bad weather. When the race was run the sky was clear and the track, in spite of the rain, was in fair condition. The time, 2:08 1/5, was considered remarkable." — *Courier-Journal*

"Over a rather heavy track . . . Cold weather and showers which fell until noon, kept the attendance down to 15,000." — *Times-Picayune*

Rainfall on Derby Day amounted to 0.21 inch. Maximum temperature was 57°F (14°C).

May 11, 1918

Exterminator came from behind to carry off this one in 2:10 4/5.

"After raining all morning, the skies cleared for the Derby. After the races were over, the downpour began again and continued into the night . . . The Derby of 1918, due to weather conditions, will always rank with the victories of such despised plugs as Stonestreet, Pink Star, and Donerail . . . through mud and slush." — *Courier-Journal*

Exterminator was a 30-to-1 long shot, paying $61.20 to win.

Rain fell very heavily all day to the substantial total of 2.31 inches. The maximum temperature was 68°F (20°C).

May 16, 1925

Flying Ebony really flew despite a sloppy track, winning in 2:07 3/5.

"Immortality descended on a horse named Flying Ebony between two thundershowers at Churchill Downs yesterday afternoon.

"Thousands, dampened, but too excited to be daunted by a shower that preceded the classic, were still rapturously cheering the little black colt's victory when a second downpour, decidedly more emphatic than the first, drenched the largest throng that ever saw a Kentucky Derby . . .

"The two showers followed a flawless day. They came just in time for the Derby, and just in time to save the Weather Bureau's name for accuracy, for rain had been forecast. Post time of the $50,000 race was advanced after thunder and lightning . . .

"The first four races went smoothly, while the operators watched the clambering gray clouds. When the Derby post time was moved up, the first patter of rain had sprayed the windows. When the storm broke it nearly tore out the equipment, while the hail shower blurred the windows until it was almost impossible to tell accurately which horse was in the lead." — *Courier-Journal*

The day's rainfall amounted to 0.14 inch. The maximum temperature was 83°F (28°C).

May 19, 1928
Days of rain and a heavy track made Reigh Count the favorite. He won in 2:10 2/5.

"The track, beaten and flooded by heavy rain for three straight days, was further dampened by a downpour which began just before the fourth race . . . At 3:57 a drizzle began and a moment later it developed into a downpour . . . The track, which was already in a miserable condition, was worse . . . Through deep puddles the starters picked their way." — *Courier-Journal*

"Some 80,000 frenzied persons, many of them drenched to the skin as they stood in the beating rain that came a half hour before the race . . . struggling animals enmeshed in the watery mud that covered the racing strip . . . It did not rain during the encounter, but it rained hardly a half hour before, and a course that was already muddy and slimy was made into a rippling sea of oozy mud that splattered horses and riders and almost silenced the ringing hoof beats." — *New York Times*

Three days of moderate-to-heavy rain preceded the running of the race: 2.77 inches on the seventeenth; 0.57 inch on the eighteenth; and 0.62 inch on the nineteenth. The maximum temperature on Derby Day was 71°F (22°C).

The wettest Derby on record, Churchill Downs, 1929. Reprinted with permission from the Courier-Journal *and* The Louisville Times.

May 18, 1929

Clyde Van Dusen won in 2:10 4/5 on a muddy track.

"During most of the afternoon it rained, and for the second straight time the Derby was run over a sloppy track . . . At 4:38 shortly before the bugle called the field to the post, a driving rain that could have made it impossible to separate the horses had the race been run in it set in." — *Courier-Journal*

"A drenched throng of 75,000 persons from all parts of the country withstood a rainstorm that burst with tropical fury an hour before post time and left the track thicker than an asphalt pool in Trinidad . . . Water became inches deep in the Clubhouse . . . following luncheon when the sun came out . . . but more showers fell." — *New York Times*

The rainfall was heavy, amounting to 1.19 inches at the downtown weather bureau station. The maximum temperature was 77°F (25°C).

June 9, 1945

This day was known as the "Street Car Derby" as a result of automobile use restrictions imposed as a war measure by the Office of Defense Transportation. The track was muddy, but Hoop Jr. won in 2:07.

"Running over a track soaked by the three days of rain and under overcast skies that threatened all day long to drench the crowd of 65,000 . . . There had been a miniature cloudburst earlier . . . The sun finally peeped through even though most of the people had lost hope. It continued to play hide and seek behind the thinning mist for the remainder of the day." — *New York Times*

Three days with rain preceded the running of the Derby: 0.49 inch on the seventh; 1.00 inch on the eighth; and 0.50 inch on the ninth. The maximum temperature on Derby Day was only 51°F (11°C).

May 3, 1958

Tim Tam won in 2:05 on a muddy track.

"Despite cloudy skies that showed little signs of clearing up . . . the weather grew pleasant late in the day and the sun broke through intermittently." — *New York Times*

"Rain Friday and a forecast of more yesterday kept race goers from nearby communities away. As it turned out, it didn't rain a drop, and there were sunny interludes in the clear sky. The weather was humid and muggy and maybe that didn't help Silky . . .

"Umbrellas became parasols yesterday as the clouds vanished. Raincoats were used to provide shade." — *Courier-Journal*

Rainfall amounted to 0.30 inch on the second, and to 0.17 inch on the third. The maximum temperature soared to 85°F (29°C) on Derby Day.

*   *   *

Since 1958, no Kentucky Derby race has been run under track conditions that were listed as unfavorable, nor have adverse weather conditions been present during the race. On May 7, 1983, however, a thundershower about forty minutes before post time sent many of the audience scurrying for shelter, but so efficient was the drainage system that the track was in good shape for the running of the race.

Boston Marathon

"This is the one simon-pure athletic event left in the United States. The Boston Marathon draws more people than the Kentucky Derby or the Indianapolis 500 and it doesn't cost anyone a cent to see it. This is the charm of the old race. That it is so devoid of commercialism." So wrote sports historian Joe Falls in *The Boston Marathon*.

The idea of having a long-distance footrace in the Boston area sprang from the experiences of some members of the Boston Athletic Association who attended the revival of the Olympic Games in Greece during the summer of 1896. The marathon was the feature of the games, so a similar race was scheduled for Patriots Day, April 19, 1897, when the B.A.A. annually staged a track meet in Boston.

The first race of approximately 25 miles started in the town of Ashland, west of Boston, and finished with a lap around the track at the B.A.A.'s Irvington Street Oval. It was a "beautiful April day" with the temperature rising to 69°F (21°C) at Framingham, near the start of the race, and to 66°F (19°C) in downtown Boston. New England lay in the warm sector of a storm system; with the passage of its cold front the next morning the temperature plummeted to well below freezing.

The race has always been held on or close to April 19, in the ficklest part of the ficklest month of the year in eastern Massachusetts. It can belong to winter or to summer, as the extreme temperatures experienced in Boston after April 15 well demonstrate: 21°F (−6°C) on April 20, 1897 (the day after the first marathon), and 94°F (34°C) on April 18, 1976 (the day before the 1976 marathon).

At this season, the waters of Massachusetts Bay are still close to their lowest temperature of the winter, while inland, with the snow cover gone, the bare earth heats rapidly under the rays of a sun mounting higher and higher each day. Rising currents of warm air over the land during the day entice a cool sea breeze to flow landward. This circulation, known locally as a "sea turn," often sets in during late morning or early afternoon and works its way inland gradually. At some time in the early afternoon the Wellesley Hills may mark the advance front of the westward-moving cool air.

Thus, the Marathon at its starting point in Hopkinton may experience torrid conditions of a warm spring day, but the race may end under the influence of a cool sea breeze in downtown Boston. This was the situation on April 19, 1976, when gas station thermometers at Hopkinton were at 100°F (38°C) at the start, but the runners finished in 68°F (20°C) air just two hours later.

The race starts in Hopkinton on Hayden Rowe in front of the house of Mrs. Frances McMannus. It proceeds on Route 135 (Central Street) through Ashland, Framingham, and Natick to Wellesley. The course turns northeast on Route 16 (Washington Street) to Commonwealth Avenue in Auburndale (Route 30); then eastward through the rest of the Newtons on Route 30, over the four heartbreak hills near Boston College, and past the Chestnut Hill Reservoir to Cleveland Circle. The approach to downtown Boston is by Beacon Street to Kenmore Square, where Commonwealth Avenue is met again. The race ends at the Prudential Center, which lies two blocks south of Commonwealth Avenue, facing Boylston Street. The distance from start to finish is 26 miles and 385 yards (42.19 km), the distance of the Olympic Marathon in 1908 and standard since 1924.

Thermometer readings in the Boston area must be used with caution. Along the route of the Marathon the temperature at the National Weather Service substations at Framingham and Chestnut Hill may be employed for good approximations of conditions for the first and middle portions of the race course. These are taken in standard instrument shelters with the thermometer in the shade and ventilated by whatever wind is blowing.

Until January 1, 1936, the official Boston weather station was located downtown in the vicinity of the U.S. Post Office and U.S. Custom House in the financial district, where thermometers were located on rooftop exposures, and thus were not representative of street-level conditions. With the movement of the instruments to Boston Airport (now Logan International Airport) in East Boston in 1936, the official readings became even less representative of conditions along the city portion of the Marathon course since the airfield juts out on built-up land into the waters of Boston Harbor, which are still cool in late April. The finish line at the Prudential Center is located about 3.75 miles from the airport station.

Accordingly, the readings at Framingham and Chestnut Hill are often more representative of race conditions than those reported for Boston. Perhaps the "gas station thermometers" that some reporters choose to quote are really the best indication of the thermal condition of the asphalt race course and of the runners thereupon.

The coldest Marathon Day came in 1925, when the temperature at downtown Boston ranged from 32°F (0°C) in the morning to 38°F (3°C) in the early afternoon. A steady north wind of about 20 mi/h (32 km/h) made the wind chill even more penetrating. The second coldest came in 1967, when both Framingham and Chestnut Hill reported maximums of only 38°F (3°C) with a drizzling rain falling.

There is no doubt that Marathon Day in 1976 was the hottest of all. Temperatures that day in Massachusetts reached 100°F (38°C) on the most remarkable early-season hot day in the entire New England record. Framingham reported an official 94°F (34°C). The hot air did not penetrate to Boston until the race was over, but next day the airport thermometer soared to 94°F (34°C), the hottest reading until May 9 in the entire Boston record since 1871.

The chances of precipitation on an April day in Boston are one in three, but since the race requires only about three hours, the chances of rain or snow falling on the runners in action is much less. The Marathon has been lucky in not experiencing the driving rains and winds of a major northeaster. The first half of the 1912 race was run under rainy conditions as a storm was ending. The race in 1939 was probably the worst from a rain point of view; a total of 0.64 inch fell during the daylight hours, slowing the runners' time and cutting down the crowds considerably. On many occasions a northeast storm has threatened with low clouds and cool temperatures but without rain, conditions greatly favored by the runners.

The chances of snow by Patriots Day have dropped to less than one in thirty. In fact, only four Marathons have seen snowflakes in the air, and none has been started with measurable snow on the roads. But the snow season has not vanished entirely. The latest date that snow has whitened the ground in downtown Boston was on May 9–10, and that occurred as recently as 1977, when all central New England had a record late snowstorm. This was the first measurable snowfall in May in the official Boston records that date back to 1871. Other late-season measurable snows have been 3 inches on April 24, 1874, and 4 inches on April 18, 1887. The granddaddy of all late April snowstorms struck on April 17–18, 1821, when many eastern Massachusetts towns reported 12 to 15 inches. So much snow clogged the roads that the Massachusetts legislature could not muster a quorum and had to postpone the opening day. Afternoon temperatures in downtown Boston on the seventeenth were just 32°F (0°C) and on the eighteenth 39°F (4°C). On Patriots Day the mercury climbed to 44°F (7°C), so the roads would have been filled with a foot of slush, and a race of any length would have been out of the question. Remember, it could happen again!

In the following section, weather conditions are described for some of the Marathons when they were somewhat adverse for the runners.

1909: Scorching

"In no year since the B.A.A. started this great contest have the conditions been so unfavorable. On two other occasions the day has been warm, but yesterday the sun's rays were scorching . . . To brave a thermometer of 80 in the shade was something they had not dreamed of . . . they found they had not the reserve force to hold the pace under the broiling sun, and many a famous runner was forced to quit." — *Boston Transcript*

The maximum temperature at Framingham was 82°F (28°C), at Chestnut Hill 85°F (29°C), and at Boston 83°F (28°C).

Henri Renaud of Nashua, New Hampshire, was the winner in the slow time of 2h 53m 36s.

1912: Rain and Mud

"The race was run in a rainstorm from the start and which continued until the reservoir was reached. The roads were inches deep in mud and the runners were bespattered from head to foot. The road conditions were about the worst in the history of the famous race, although the atmospheric conditions could hardly have been better. The runners however took to the sidewalks whenever they were available, and this saved them a great deal of effort, especially on the heart-breaking hills . . . The road was not at its best today because of the northeaster, but the air was cool and what breeze was stirring came across the course." — *Transcript*

Rain fell every day from the fourteenth through the nineteenth; on the eighteenth the amount was 0.40 inch and on the nineteenth, 0.11 inch. The maximum temperature at Boston was 51°F (11°C).

The winner was Mike Ryan of New York City in 2h 21m 18s.

1915: Oppressive

"The weather was so warm and the weaker athletes began to fall by the wayside, many of them complaining of the dust and the gasoline fumes. There was scarcely a breath of air, and the heat was more oppressive than in former years." — *Transcript*

The maximum temperature in Boston was 72°F (22°C). There were press reports of 84°F (29°C) along the course.

Edouard Fabre of Montreal, Province of Quebec, was the winner in 2h 31m 41s.

1925: Snow and Cold

"It was one of the bitterest days the race ever saw. The temperature was close to the freezing point and a biting northwest blast hitting square in the faces of the runners all the way from start to finish did a lot to discourage. To cap everything the boys ran through a young blizzard through the first hour of the journey. The snow that fell after the start at Hopkinton may not have left any traces, but the way it whipped into the faces and the thinly clad bodies of the athletes was heartrending.

"We, covered in our automobiles, were cold and stung by the snow blast. What must the athletes have suffered. I saw a number finish with numbed arms and hands. They had to be assisted in getting their clothes off and they had to take hot shower baths before they were able to get feeling enough into their limbs to handle the food offered them." — *Herald*

The maximum temperature at Boston was 38°F (3°C) and the minimum 32°F (0°C). The maximum wind was from the north at 22 mi/h (35 km/h), corrected.

The winner was Charles Mellor of Chicago, beating Clarence De Mar by 33 seconds, in 2h 25m 40s.

1939: Downpour

"Despite weather conditions which were the worst since 1912, when Mike Ryan laughed off a similar storm . . . The rain, which started gently when the runners had covered about seven miles, poured down during the latter stages of the race, reducing the crowd to perhaps 100,000, one of the smallest turnouts in the history of the race . . . While the downpour discouraged the spectators, it served to refresh the runners and spur them on to their best performances. The only possible deterrent developed in the form of slippery roads, but few of the leaders were bothered." — *Herald*

The Boston Airport temperature at noon was 43°F (6°C), the relative humidity 96 percent, and the wind east at 8 mi/h (13 km/h). A total of 0.64 inch of rain fell from 8:00 A.M. to 8:00 P.M.

The winner was Ellison M. (Tarzan) Brown of Alton, Rhode Island, in 2h 28m 52s.

1941: Warm

"A scorching sun turned the highway from Hopkinton to Boston into one huge, red-hot stove cover, but Pawson spurned the blistering pavement . . . The thermometer hit a new high and the starting field hit a new low in the 45th B.A.A. marathon yesterday. Fortunately the two did not coincide; the thermometer was 74°, the starting list 124. Wouldn't it have been terrible if they had been reversed . . . It was plenty hot, however. In fact, it was so hot that the runners appeared to be running in cellophane panties. They doused themselves with water and practically had a coming out party." — *Herald*

The airport temperature was up to 75°F (24°C) during the afternoon.

The winner was Leslie Pawson of Pawtucket, Rhode Island, in 2h 30m 38s.

1964: Wet

"Wet snowflakes plastered the record number of 301 starters at Hopkinton and a cold, wet mist saturated them all the way to Natick. The temperature was a frosty-breathed 39 degrees and the sharp edges of a head wind shaved the athletes throughout." — *Globe*

The maximum temperature at Framingham was only 43°F (6°C), at Chestnut Hill 50°F (10°C), and at Logan Airport 47°F (8°C). Precipitation at Boston was 0.04 inch.

The winner was Aurele Vandendriessche of Belgium in 2h 19m 59s.

1970: Wet

"Ron Hill, first Briton to capture the Boston A.A. Marathon in 74 years, came swooping through the torrents of rain yesterday, graceful as a herring gull . . . They splashed over the final 10 miles almost within reach of each other . . . Yesterday's marathon crowd was more sparse than at any time in recent years because of the weather.

"Pat Paulsen, a woman contestant, said: 'Only in Boston could you get so many crazy people out freezing to death and dying of pneumonia. I guess it's the charm of the city.' " — *Globe*

The maximum temperature at Framingham was 58°F (14°C), but at Logan Airport it was only 46°F (8°C). Boston's Logan Airport measured 0.20 inch on the twentieth.

1976: The Hottest Race

"Through the first half of the race the weather was brutal and the humidity was such that a person could work up a sweat just by standing around. Fortunately people lining the course used their garden hoses to cool off the runners.

"By the time the runners got to Wellesley a sea breeze had come up, easing the heat considerably. From the halfway point the temperature dropped drastically, until the thermometer read only 68 degrees when Fultz crossed the finish line." — *New York Times*

"The first of the four deadly climbs that were even more perilous than ever on this, the hottest day ever for the Marathon. The temperature was 100 degrees at the start at Hopkinton . . . Two gas station thermometers at Hopkinton read 100 and 101 at 11:30 A.M." — *Globe*

Official temperatures in thermometer shelters were reported as high as 100°F (38°C) in Massachusetts on April 19, 1976. It was not only the hottest April day in history, but also exceeded the hottest day ever experienced in May at many locations. Framingham reported an official 94°F (34°C), Chestnut Hill 91°F (33°C). Logan Airport in East Boston had been up to an all-time April record of 94°F (34°C) on the day before, the eighteenth, but a sea breeze kept the temperature down below 70°F (21°C) on the nineteenth until after 4:00 P.M., when the race was over for most of the participants.

Jack Fultz of Franklin, Pennsylvania, was the winner in 2h 20m 19s.

Weather Notes on Recent Marathons

1978, April 17 — "an ideal 46-degree day."

1979, April 16 — "a drizzle-drenched course . . . it was so cold." Boston's temperature range was only from 45°F (7°C) to 40°F (4°C).

1980, April 21 — "a hot day" — "road level in 80s under an ardent sun although a low humidity." Boston's maximum was 71°F (22°C), Framingham's 74°F (23°C).

1981, April 20 — "a cool day" — "perfect 49-degree weather." Boston's maximum was 53°F (12°C).

1982, April 19 — "a warmish day in direct sunshine" — "It was on a 68-degree grill made warmer by unclouded sunshine." Boston's maximum was 70°F (21°C).

1983, April 18 — "a cool, cloudy day" with "an early tailwind and 40-degree weather," according to the local press. These were ideal conditions. Boston weather station reported maximum afternoon reading of 53°F (12°C) and Framingham reading was 52°F (11°C). No rain fell during the race.

CHAPTER FIVE

The Realm of Flight

Air in motion both lifts and propels an aircraft upward and forward, and current atmospheric conditions are all-important in determining whether an aerial vehicle will have an easy takeoff, a smooth flight, and a safe landing. Not until man essayed to travel through the air did the science of meteorology achieve the dominant position in travel planning that it holds today.

A cool, calm day enabled the first American balloon flight to be launched and completed successfully at Philadelphia in 1793. If a stiff northeast wind had not been blowing, it is quite possible the Wright Brothers might not have been able to get their craft into the air on that historic day at Kitty Hawk. An encounter with an adverse wind frustrated the first dirigible flight toward Europe in 1910 when the *America* was well on her way on the projected flight path.

Of the pioneer transatlantic flights by heavier-than-air planes, those attaining success encountered tolerable weather, while many that failed met adverse conditions bringing disaster. The first airmail flights were subject to the twin adversaries of weather and mechanical failure in about equal proportions. The loss of all three of the U.S. Navy's principal airships, and possibly that of the *Hindenburg*, was attributable to encounters with unfavorable weather situations. In recent years, space flights have been postponed until visibilities and ceilings were such that the launch could be tracked visually as well as electronically. Anything that flies through the air needs favorable atmospheric conditions.

Aviation Firsts

First American Aeronauts

Upon reading of the exploits of the Montgolfier brothers in demonstrating the first free balloon flight at Annonay in the Rhone Valley of France on June 4, 1783, Thomas Jefferson gave the subject much thought. In a letter from Annapolis on April 28, 1784, he enumerated a number of uses for aerial transport in the future. In addition to its role in carrying men and intelligence from place to place, he envisaged the balloon's employment in "throwing new lights on the thermometer, barometer, hygrometer, rain, snow, hail, wind and other phenomena of which the Atmosphere is the theatre." Benjamin Franklin, a witness of a Paris ascent of a Montgolfier balloon, was quick to see the potentialities of flight and urged members of the American Philosophical Society at Philadelphia to pursue the new field of scientific enterprise.

The first American to make an ascension in a free balloon was Dr. John Jeffries, a Boston-born physician who was residing in London following his participation in the Revolutionary War as a surgeon with the British forces. Long an avowed student of meteorology, Jeffries had maintained a weather diary while practicing medicine in Boston before the war. (Parts of the diary have been preserved and may be consulted in the archives of Harvard University.) Jeffries perceived the balloon's utility as an adjunct of his meteorological studies. Among its many uses, he foresaw that "it could more clearly determine heights from the earth, and . . . by observing the varying course of the currents of the air, or winds, at certain elevations, . . . throw some new light on the theory of winds in general."

In order to satisfy his curiosity, Jeffries paid for permission from Jean Pierre Blanchard, a pioneer French aeronaut, to participate in a demonstration flight over London and environs. On November 30, 1784, he made an aerial voyage to Kent lasting one hour and twenty-one minutes during which he took frequent readings of his barometer, thermometer, and other scientific instruments. Jeffries also participated on January 7, 1785, in the celebrated first cross-Channel flight with Blanchard from Dover to the Forest of Guines, 12 miles inland from Calais. The Anglo-American wrote of his experiences in *A Narrative of Two Aerial Voyages of Doctor Jeffries with Mons. Blanchard, with Meteorological Observations and Remarks,* published in London in 1786.

The distinction of being the first to lift off American soil in a balloon flight has been clouded in doubt as a result of a hoax perpetrated by a letter of unknown authorship that appeared in the *Journal of Paris,* supposedly written at Philadelphia on December 29, 1783. Jeremiah Milbank, Jr., in his *The First Century of Flight in America* has found no evidence that such a flight took place either in contemporary letters or diaries, nor does the local press make mention of any such exploit.

First Free Balloon Flight in America — January 9, 1793

During the turmoil attending the outbreak of the French Revolution, Jean Pierre Blanchard brought his balloon to Philadelphia, where he made the first professional flight in the New World. On January 9, 1793, in the presence of President George Washington and Mr. Ternan, the French Minister Plenipotentiary, the aeronaut rose from the Prison Court at the southeast corner of Sixth and Chestnut streets and was carried first southeast, then south-southeast to a point a little east of Woodbury across the Delaware River in New Jersey, an airline distance of about nine miles. The aerial voyage lasted for forty-six minutes.

The weather on this January morning was most favorable for the enterprise: clear, cold, and almost calm. In his published account of the trip, *Journal of My Forty-Fifth Ascension, being the first performed in America, on the ninth of January, 1793,* Blanchard reported the following meteorological conditions:

> At 9 minutes after 10, the sky being clear, serene and propitious, little wind and nearly calm at the surface of the earth; Reaumur's thermometer in the sun $10^d 5/10$, Fahrenheit's $55^d 6/10;$* corrected

---

*The free air temperature, probably in the shade, at 9:45 A.M. had been 3° Reaumur (38.8°F).

altitude of the barometer 29 inches, 7 lines 2/10, English measure, I affixed to the aerostat my car, laden with ballast, meteorological instruments, and some refreshments, with which the anxiety of my friends had provided me. I hastened to take leave of the President, and of Mr. Ternan . . .

Blanchard mentioned that the two men holding the gasbag were Messieurs Nassy and Legaux, the latter being a French émigré who possessed the most elaborate set of meteorological instruments in America at that time and made daily observations for many years at Spring Mill, about 13 miles northwest of Philadelphia.

After twenty-nine minutes of flight, the balloon had risen to an altitude of over one mile. Blanchard described the moment:

> The lowest state of the mercury in the barometer after having brought its surface in its lower reservoir to its proper level and corrected its dilation, was 69 lines 9/16 French measure, or 74 lines 8/16 English measure, which according to Mariot, Boyle, Deluc and Father Cote gives an elevation of 905 toises 1 foot 6 inches (the toise at 6 feet) or 5431 feet 6 inches French measure, and at the usual reduction 968 fathom 4 feet, or 5812 feet English measure.

Blanchard's takeoff from Philadelphia on the first balloon flight made in America, January 1793. Smithsonian Institution Photo No. 32619E.

This was the highest elevation of my balloon, without having thrown out any of my ballast, except the liquor contained in the 6 bottles given to me by Doctor Wistar.

At this moment (10^h 38^m) the thermometer of Reaumur 9^d Fahrenheit's 52^d 3/10 (the temperature of the air most delightful and quite extraordinary for this season of the year). These observations were made with so much the more confidence, as I enjoyed for a long time the calmest reflection.

In the mean while the state of the atmosphere began to change. A whitish cloud withheld from my sight for several minutes a part of the city of Philadelphia, which appeared to me only as a most minute and microscopic object. A thick fog covered the south; towards the east, in the lower region of the atmosphere, a mist arose, which prevented me from reconnoitring the sea. I was afraid that the land-wind encreasing, might render my descent difficult; every thing induced me to accelerate it.

First Heavier-Than-Air Flight:
Kitty Hawk — December 17, 1903

Kitty Hawk lies in historic country close to remote Roanoke Island on the Outer Banks of North Carolina. The principal topographic feature of the area is Kill Devil Hill, rising to an altitude of 91 feet above the sandy beaches. It was the presence of this eminence close to the seashore, where dependable sea breezes usually prevail, that brought the bicycle-building brothers, Orville and Wilbur Wright, from Dayton, Ohio, to the Carolina coast to test their gliders. In response to a request for information, they received word from Joseph J. Dosher, the official-in-charge of the U.S. Weather Bureau station at Kitty Hawk:

> In reply to yours of the 3rd, I will say the beach here is about one mile wide, clear of trees or high hills and extends for nearly sixty miles same condition. The wind blows mostly from the north and northeast Sept. and October . . . I am sorry to say that you could not rent a house here, so you will have to bring tents. You could obtain board.

Starting in 1900, the Wrights came to Kitty Hawk for three successive vacations in late summer and autumn to test the aerodynamics of their man-carrying gliders. When the brothers returned in the autumn of 1903, they brought a glider-type craft equipped with a newly developed, lightweight, gasoline-fired engine. Tests were made on several aspects of the planned flight, and all was ready by dawn of December 17, 1903, when the early morning wind and weather conditions were deemed suitable for flight.

The weather map at 8:00 A.M. on Thursday, December 17, showed a strong anticyclone at 30.45 inches (103.1 kPa) over Wisconsin with a ridge of high pressure bulging east to the Delmarva peninsula on the Atlantic Coast. This placed Kitty Hawk in a northerly air flow. Norfolk reported north at force 4 (13 to 18 mi/h or 21 to 30 km/h), and Kitty Hawk had the same speed at northeast. The sky was partly cloudy, and the temperature read 28°F (−2°C) at Norfolk, and cloudy conditions prevailed at Kitty Hawk with the mercury at 38°F (3°C).

The Wright brothers described the local wind behavior that morning in a subsequent joint report of their activities:

> We had a "Richard" hand anemometer with which we measured the velocity of the wind. Measurements made just before starting the first flight showed velocities of 11 to 12 meters per second, or 24 to 27 miles per hour. Measurements made just before the second flight gave between 9 and 10 meters per second. One made just afterward showed a little over 8 meters. The records of the Government Weather Bureau at Kitty Hawk gave the velocity of the wind between the hours of 10:30 and 12 o'clock, the time during which the four flights were made, as averaging 27 miles at the time of the first flight and 24 miles at the time of the last.

(Tests made in the 1920s showed that these early anemometers gave speeds too high; corrections for the latter two figures would be 22.7 and 20.5 miles per hour.)

Orville Wright's diary gave further details on the wind conditions and their effect on the historic events of December 17:

> When we got up, a wind of between 20 and 25 miles was blowing from the north. We got the machine out early and put out the signal for the men at the station ... After running the engine and propellers for a few minutes to get them in working order, I got on the machine at 10:35 for the first trial. The wind according to our anemometer [a Richard instrument made in Paris] at this time was blowing a little over 20 miles (corrected) 27 miles according to the Government anemometer at Kitty Hawk.

Orville also related that the biting cold in the stiff wind made the work difficult and several times forced them to retire to their nearby shack, where a good fire was burning in an improvised stove made of a large carbide can. December 1903 was a cold month on the Outer Banks, with the temperature averaging 5°F (3°C) below normal. Later the waters of Albemarle Sound froze over more than a mile from shore, an unusual event.

Four flights were made between 10:30 A.M. and 12 noon. The fourth and last proved the most successful by far, carrying through a distance of 852 feet during an airborne glide of 59 seconds. At the end, winds gusting over a hummock brought the plane down with some damage to the front rudder, but the main frame suffered little. However, while the aviators and assistants were standing around and discussing the results of the last flight, a gust picked up the entire machine and rolled it over. By the time they were able to fasten it down, the legs supporting the engine were broken off and parts of the ribs and a spar were badly damaged. They were forced to abandon a planned attempt to fly from Kill Devil Hill to the weather station at Kitty Hawk four miles distant.

Orville Wright sent the first news of the historic adventure to his father at Dayton:

KITTY HAWK, DECEMBER 17, 1903

SUCCESS FOUR FLIGHTS THURSDAY MORNING ALL AGAINST TWENTY-ONE MILE WIND STARTED FROM LEVEL WITH ENGINE POWER ALONE AVERAGE SPEED THROUGH AIR THIRTY-ONE MILES LONGEST 57 SECONDS INFORM PRESS HOME CHRISTMAS.

In addition to the error in the telegram, giving 57 seconds instead of 59 as the gliding time, many erroneous reports circulated in the press about the exploit. To counteract these, the Wrights released a statement as to the correct nature of their flights. This was published in the December 1903 issue of the *Monthly Weather Review* along with a brief statement from weatherman Dosher:

The Weather at Kitty Hawk — Their success is undoubtedly due in great part to the preliminary careful study of the winds, and for this reason, although machinery is essential, yet we consider that meteorology also has played an important part in their work. Their official announcement is that on December 17 four flights were made. The wind record at Kitty Hawk was 24 to 27 miles per hour at 30 feet above the ground.

These figures should be corrected to 20.5 to 22.7 mi/h (33 to 36.5 km/h), in accordance with later calibration corrections applied to all weather bureau anemometers then in use.

First Airplane Flight Across the Atlantic: NC-4 — May 1919

"Early in September, 1917, Rear-Admiral D. W. Taylor called Naval Constructor Hunsaker and me into his office, and took our breaths away by giving us terse instructions to begin and design a seaplane that could cross the Atlantic under its own power." This is the opening paragraph in Commander G. C. Westervelt's *The Triumph of the N.C's*, the story of how one of three flying boats succeeded in becoming the first aircraft to cross the Atlantic Ocean. NC was a contraction for Navy-Curtiss, the latter for Glenn Curtiss, the leading American aircraft designer and manufacturer.

At the end of World War I in November 1918, *NC-1* alone had performed a trial flight; the others were still being fabricated. Interest in a transatlantic flight, however, was soon rekindled by the renewal of Lord Northcliffe's *Daily Mail* prize of £10,000 (about $50,000) to "the first aviator who shall first cross the Atlantic (either way) in an aeroplane from any point in the United States, Canada, or Newfoundland to any point in Great Britain or Ireland, in 72 consecutive hours." In addition to the showmanship, the purpose was "to stimulate the production of more powerful engines and more suitable aircraft."

Though the U.S. Navy was not technically qualified for the conditions of the contest, it joined in the spirit of the times and prepared for a flight to take place in the spring, when weather conditions were judged to be favorable for such a venture. *NC-3* and *NC-4* were rushed to completion, and in this process *NC-2* fell victim to cannibalization of its parts for the benefit of the others.

At least six other aviation groups were either at or en route to Newfoundland in early May 1919 when the seaplane squadron assembled at Rockaway Beach Naval Air Station, located at the most seaward point of New York City between Jamaica Bay and the Atlantic Ocean. The three planes were commissioned on May 2 while work was still in progress to fit them for their initial trials. *NC-4* became known as the "lame duck" of the outfit as a result of a series of misadventures, and this reputation won it a sort of affection with the American public, who were kept informed of the preparations through wide press coverage.

All was set for the takeoff on May 7 toward the northeast along the route to Halifax in Nova Scotia and Trepassey Bay in Newfoundland, but the weather did not cooperate, so the departure was delayed a day. Finally, after discouraging reports in the early morning of May 8, conditions began to improve along the route as dry air from an anticyclone over the Great Lakes replaced the moist, maritime circulation. The takeoff was scheduled for 10:00 A.M.

The attending conditions were well described by a reporter of the *New York Tribune*:

> At the moment "four bells" rang clear above the engine noise the NC-3, flagship of Commander John H. Towers, "admiral" of the transatlantic seaplanes, started on the first leg of her momentous aerial voyage. With a roar from the concentrated power of 16,000 horses the huge ship tore through the haze-swept waters of Jamaica Bay. Fifty seconds later she literally leaped from the water into the air at terrific speed.
>
> The haze had scarcely shadowed her graceful lines when Lieutenant Commander A. C. Read's NC-4 rushed through the water after her. The "Four," too, rose with graceful ease, and quickly pierced the misty shroud.
>
> Then the NC-1, first of her type, and commanded by P. N. L. Bellinger, raced madly after her two sisters. Sixty seconds after her start she sprang into the air.

Not all the planes made it to Halifax by night. Lame duck *NC-4* developed an oil leak and had to put down in the ocean when some 80 miles off the tip of Cape Cod. Unable to raise assistance from nearby craft, *NC-4* started to taxi over the water toward land. "It soon got dark, but it was a beautiful night, with the moon nearly full, and the sea had smoothed down, so that after the first hour or two we made about ten knots. We were not much worried, for if the engines held out we were bound to make shore. They did hold out all through the night," wrote Lieutenant Commander Albert C. Read, skipper of NC-4, which made the harbor at Chatham Naval Station without aid at about daylight. They were lucky in their timing, for the next morning a three-day gale arose and kept them tied down in the harbor.

Eventually, after another engine mishap, NC-4 arrived at the Newfoundland base expecting to find NC-1 and NC-3 gone, but they were still there when the laggard plane put down on the afternoon of May 15. The other two seaplanes had attempted to leave but were unable to get off the water in view of their over-weight and crosswinds. Next day, loads were lightened and on a second try the three aircraft became airborne and headed for Ponta Delgada in the Azores Islands, some 1350 nautical miles distant. A sizable fleet numbering over sixty cruisers, destroyers, and smaller vessels was strung out over the route for navigational assistance and rescue operations if necessary.

Commander Read described the forecast situation as relayed to him from Lieutenant Commander Alexander McAdie, peace-time professor and superintendent of the Blue Hill Meteorological Observatory of Harvard University, who was aboard the cruiser *Baltimore*:

> But the aerographic officer — in plain English, the weather fore-caster — informed us that conditions all the way to the Azores would be unusually good, and that a change would probably occur if the start were postponed for another twenty-four hours. He was right, except that, as will be seen later on, the changes came sooner than expected — a storm sneaked in from a quarter unguarded by any one of our "meteorological battleships."

The reconstructed weather map for 1:00 P.M., GMT, May 16, or 9:00 A.M., Newfoundland time, when the decision to take off was made that afternoon, placed a mature frontal system over the central Atlantic Ocean along longitude 40°W, running from about 55°N latitude southward to an occlusion point, where the cold front and warm front joined, at 43°N and 42°W. This location was close to the route that the three planes would fly, about one-third or 450 nautical miles from Trepassey Bay to the Azores. The entire frontal system moved east at a rate of about 35 mi/h (56 km/h) during the next twenty-four hours, according to comparative map positions. This meant the planes would catch up with the frontal zone soon after dawn the next morning.

Commander H. C. Richardson, pilot of *NC-3*, described the conditions met by the three planes after takeoff and during the night while they remained in sight of each other. The air was rough at the harbor entrance as they climbed, and it continued bumpy, especially in the vicinity of icebergs. The sun set in a heavy overcast. "Throughout the night the *NC-3* found the air restless, and we tried different altitudes searching for better conditions," he wrote, "but to little avail." It became difficult to maintain course and balance and required the attention of two pilots constantly. The aircraft were in an unstable polar airstream, a northwesterly flow whose main characteristic is turbulence. "Very few stars were visible, and until the moon was well up we had to depend principally on the instruments," Richardson continued. Air conditions remained rough most of the night until about 4:00 A.M., when the ship climbed above the clouds and "found smooth air for the first time." But soon another menace, in the form of fog and poor visibility, appeared as the flight approached the occluded frontal system.

At this time *NC-4* lost contact with the others, so we shall follow her course during the next few hours. It was close to 8:00 A.M. when Lieutenant Commander Read's plane first began to encounter fog and rain in the frontal system that had moved eastward during the past twenty-four hours to a position approximately 180 nautical miles northwest of the westernmost Azores. Three destroyers in this vicinity reported overcast skies, wind from the west at force 5 (19–24 mi/h or 31–39 km/h), and the temperature just above 60°F (16°C). All three planes, if on a direct course, would have flown very near the point of occlusion of the frontal system, an area where flying conditions are usually the most unfavorable.

Read has retold the story of the morning's flight when his aircraft was flying blind for three hours. In speaking of the destroyers, he wrote:

Numbers 14 and 15 were checked off in regular turn, the latter at 7:45 A.M. Ten minutes later there seemed to be a rain of considerable area ahead. Course was changed to port for a few minutes to dodge the thickest part, but we soon saw that instead of rain, light lumps of fog were forming and blowing along in the same direction we were making . . . Visibility grew less and less and we missed Number 17, although little fog was encountered until 9:40. Then it began to show us what real fog was like. At 9:45 we entered an impenetrable layer as thick as pea soup. The sun disappeared entirely.

For the next hour and a half he flew the aircraft above the fog and below a broken sky overhead with occasional patches of blue showing. "We were sandwiched in between the fog underneath and the clouds overhead. Occasional rifts appeared through which water could be seen," he wrote. Radio reports from Destroyers 19 and 20 were discouraging.

The first good news came when Number 21, some distance ahead, reported "10 miles visibility." Still thick conditions prevailed aloft, although blue sky could be seen to the south. "Then suddenly, at 11:27 or nearly three hours after passing the last destroyer sighted, while flying at 3,400 feet altitude, we looked down and on our port hand saw a tide rip through one of the rifts in the fog which had of late become less frequent." This was the western tip of Flores Island, the westernmost of the Azores. It was a most fortunate sighting and was attributable to the better visibility existing in the warm sector of the storm area, in contrast to the thick conditions in the immediate frontal zone. Two to three hours later all of the western Azores would be shrouded in low ceilings with poor visibilities along the front, and a visual sighting of the western islands would have been very chancy.

In view of the deteriorating conditions, *NC-4*'s commander decided to put down at Horta on Fayal Island, the alternate stopping place, some 150 miles short of Ponta Delgada. The landing was made at 1:12 P.M., after a flight of fifteen hours and eighteen minutes from Trepassey Bay. The average speed for the run had been 79 knots or 90 statute mi/h.

NC-1 and *NC-3* were neither so lucky nor plucky as *NC-4*. Both encountered the same thick conditions, became unsure of their bearings, and put down on the sea surface to establish radio

NC-4 completes the first transatlantic crossing, at Lisbon, Portugal, in 1919. Smithsonian Institution Photo No. 21393B.

contact with the destroyers and get a navigational fix. The seas proved rougher than anticipated and both craft were soon in trouble. *NC-1* was sighted by a Greek freighter and taken in tow until a destroyer hove in sight and took over the task. The aircraft was soon damaged by the heavy seas and abandoned. *NC-3* fought for her life for sixty hours on the surface of the sea. Her skipper exhibited excellent seamanship and navigational skill in bringing his floating craft into Ponta Delgada harbor without assistance, though the ship had been damaged by the heavy seas and was no longer airworthy.

NC-4 flew on three days later to Ponta Delgada, where a reunion with the crews of *NC-1* and *NC-3* took place. *NC-4* continued on to Lisbon on May 27 to complete the first plane trip from North America to Europe. A final leg brought her from Lisbon to Plymouth, England, the port of departure of the Pilgrims in the *Mayflower* almost 300 years before.

First Nonstop Transatlantic Crossing — June 1919

Of the four British contestants to gather in Newfoundland in spring 1919 for an attempt to win the *Daily Mail* prize for a transatlantic crossing, the *Vimy* was the last to arrive. An entry of

Vickers, Limited, it was a converted bomber, powered by two 350-horsepower Rolls-Royce engines and stripped to carry additional fuel tanks. The plane could maintain a speed of 100 miles per hour over long distances. Pilot John Alcock and navigator Arthur Whitten Brown, an American, were both former military fliers now in civilian roles in the employ of the Vickers Company.

The *Vimy* arrived in Newfoundland in a crate on May 26 and was ready for its first test flight on June 9. After another trial on June 12, it was planned to take off the next day, but a damaged landing gear caused the flight to be postponed another twenty-four hours.

The weather map was active at this time, with periodic frontal passages across Newfoundland every two or three days. On the morning of Saturday, June 14, 1919, a low-pressure center was located between Labrador and extreme southern Greenland, with a cold front extending southwest across Belle Isle Strait to the northwest of the Newfoundland west coast. St. John's in the southeast lay in the warm sector of the storm system, with southwest to west gales sweeping the island. The temperature at 9:00 A.M. was 58°F (14°C).

"Strong westerly wind. Conditions otherwise favorable" was the succinct weather report given the fliers early in the morning by Lieutenant Clemens, a Royal Air Force weather forecaster who was serving as advisor to the other contestants as well.

The approach of the cold front from the northwest caused a tightening of the pressure gradient and resultant strong air flow from a westerly quadrant. "The wind remained obstinately strong during the early afternoon, and we agreed to accept conditions as they were, and to lose no more precious time," related Brown. Takeoff plans were changed. It was decided to head in a westerly direction into the wind, and this was probably a fortunate decision since the head wind gave added lift to assist in raising the heavily laden plane into the air. Brown recalled that the wind had dropped to 30 mi/h (48 km/h) by 1:00 P.M., though a reporter pegged it at 40 mi/h (64 km/h). The plane became airborne at 1:45 P.M. local time, or 4:45 P.M. (GMT).

The forecasts of Lieutenant Clemens proved a bit optimistic, as were most advisories given to the transatlantic fliers in those days. Both weather reporting and knowledge of the intricacies of North Atlantic conditions were sadly lacking at the time. Arthur Brown has related the forecast situation:

> The meteorological officer gave me a chart showing the approximate strength and direction of the Atlantic air currents. It indicated that the high westerly wind would drop before we were a hundred miles out to sea, and that the wind velocities for the rest of the journey would not exceed twenty knots, with clear weather over the greater part of the ocean. This produced satisfactory hopes at the time; but later, when we were over mid-Atlantic, they dissolved in disappointment when the promised "clear weather" never happened.

The actual weather map of the transatlantic route at 1:30 P.M. placed the warm front from the depression to the north and northwest of Newfoundland across their route about five degrees, or 300 miles east of St. John's. It reached as far south as 45°N in midocean and then swung northeast toward Europe as a cold front to join an occluded front running north and south along about 20°W, or the longitude of western Iceland. The flight would be made in close proximity to the frontal zone, except for a brief time in midocean, when the front would be south of the projected flight route. But the main occluded front in the eastern Atlantic would be overtaken when the *Vimy* was approaching northwest Ireland, since the aircraft would be traveling about three to four times the speed of the eastward-moving weather system.

In an interview with a *Daily Mail* reporter on the evening following the morning landing, Captain Alcock related the story of the weather hazards encountered while fresh from the experience:

> We had a terrible journey. The wonder is we are here at all. We scarcely saw the sun or the moon or the stars. For hours we saw none of them. The fog was very dense, and at times we had to descend to within 300 feet of the sea.
>
> For four hours the machine was covered in a sheet of ice caused by frozen sleet; at another time the sleet was so dense that my speed indicator did not work, and for a few seconds it was very alarming.
>
> We looped the loop, I do believe, and did a very steep spiral. We did some very comic "stunts," for I have had no sense of horizon.

The winds were favourable all the way: north-west and at times southwest. We said in Newfoundland we would do the trip in 16 hours, but we never thought we should . . .

We encountered no unforeseen conditions. We did not suffer from cold or exhaustion except when looking over the side; then the sleet chewed bits out of our faces. We drank coffee and ale and ate sandwiches and chocolate.

John Alcock published a full description of the flight in the September 1919 issue of *Badminton Magazine*. This was his only full retelling of the story because he was killed before the end of the year while en route to an aviation exhibition in France. His pertinent weather comments follow:

Visibility was good at the outset, but I saw ahead the dreaded fog of Newfoundland, and within a short space we were racing between it and the clouds. For seven long hours we traveled thus, sighting neither sea nor sky. Gradually the light failed and the fog increased until presently we found ourselves buried in heavy fog. The conditions were bad but I held onto the course set by Brown till suddenly we shot into a clear patch in the clouds. Quickly Brown fixed our position from the Pole Star, Vega and the Moon, and for about half an hour we had a clear sky overhead. Then, however, thick fog enveloped us again . . . [At this juncture the plane went into a spin and pulled out only when about fifty feet from the waves.]

Getting back on to course, I climbed to 7,000 feet above sea-level. Occasionally, we saw the moon, but Brown could get no reading. I hoped to get above the fog and clouds, but it seemed impossible. Climb as we might, the clouds seemed to climb still faster. For five or six hours we continued the attempt, but in vain, hail and sleet lashing across us. At 11,000 feet we met with some reward, and the sun at last appeared trying to force its way through the clouds. We then descended, but were close over the sea before clear visibility was obtained, and we found a strong sou'wester blowing.

The break in the higher clouds that enabled Brown to get a fix on the stars occurred at 12:25 A.M. This was approximately halfway across the Atlantic in the vicinity of 30°W and must have occurred when the *Vimy* was as far north of the frontal zone, now oriented west to east, as their route was to take them. This was a fortunate break, because it enabled them to get an exact reading of their location. The sky overhead remained patchy for about an hour and a half, though the plane, itself, was flying through cloud and light fog. About 3:10 A.M., the *Vimy* ran into very thick cloud as their path was converging on the cold front, which was now angling northeast toward the main low-pressure center. For about four hours they flew through the frontal zone while experiencing all types of precipitation: snow, sleet, hail, freezing rain, and rain.

When finally east of the front at 7:20 A.M. and flying at 11,000 feet in an attempt to escape the precipitation, they were able to see the sun for a brief moment, which enabled them to fix their position as quite close to the Irish coast. The *Vimy* then descended to warmer temperatures at 1000 feet, and soon two small islands were sighted through the mist, and the aircraft crossed the main coast of Ireland at 8:25 A.M.

Their navigation had been good and soon they recognized the towers of a wireless station that turned out to be Clifden. Alcock located what appeared to be a pasture nearby and put down. However, the pasture turned out to be a soft bog. The plane upended in landing with its nose plowing into the ground, but the fliers were uninjured in the jolty landing. Rain was falling. The time was 8:40 A.M.

The *Vimy* had covered 1689 nautical miles, from the coast of Newfoundland to the coast of Ireland, in 15 hours and 57 minutes, at a rate of 105 nautical mi/h (121 statute mi/h). This constituted the first nonstop crossing of the Atlantic Ocean by air direct from North America to Europe.

First Airship Crossing of the Atlantic Ocean — July 1919

In the early summer of 1919, the German Zeppelin *L-72* was poised to cross the Atlantic Ocean in a demonstration of airship power. Since the long, cigar-shaped gas bag, measuring nearly 750 feet length fore-to-aft, had originally been designed to do just that and bomb New York City, the project was scrubbed by high Allied authorities in view of the general hostility to anything German that pervaded the Versailles Treaty period. This enabled the

R-34, a British dirigible built along the lines of a captured zeppelin, to be readied for a transatlantic flight. Elaborate plans similar to those of the U.S. Navy's venture two months before were formulated. The *R-34* cast away from East Fortune, Scotland, at 1:42 A.M., on Wednesday, July 2, 1919.

The weather prospects at takeoff were described in the publication *Log of H.M.A. R-34*, by Lieutenant General Edward M. Maitland (later Air Commander), the senior officer aboard and the representative of the Air Ministry:

> *Weather Conditions.* — At first sight the weather looks far from suitable. It is very dark — rain is falling slightly — the clouds appear to be very low, and the wind whistles mournfully round the corner of the big Airship shed.
>
> The weather reports, however, in mid-Atlantic are more or less favourable. There is a depression in the North Sea centred east of Flamboro' Head, moving slowly southward, an anti-cyclone covers the greater part of the North Atlantic, another is reported over the great lakes of Canada, whilst unusual quiet exists over the whole of the North Atlantic Ocean. The wind in the West of Scotland and North of Ireland is from the north-east, and may prove of assistance to us until we are well out into the Atlantic.
>
> Major Scott, therefore, decides, despite the bad local weather conditions, to get away as soon as possible.

British dirigible, R-34, *arriving over Long Island, New York, on first airship crossing of the Atlantic. The Telamon Collection, The University of Georgia Libraries.*

The reconstructed weather map for 1:00 P.M. (GMT), July 2, placed a barometric depression over the North Sea with a central pressure of 29.68 inches (100.5 kPa). This had caused the rain, low clouds, and northerly winds at takeoff about eleven hours earlier. A large high-pressure area covered the eastern half of the Atlantic Ocean, with its ridge line running north–south along 27.5°W latitude, or about 1000 miles west of the coast of Ireland. The *R-34* received helpful northeasterly winds from the anticyclonic system while southeast of its center, and then benefited from easterly and southeasterly flow when to the west. What was to become the major feature, though not mentioned in the briefing the evening before, was a barometric depression centered southeast of Newfoundland at 43°N, 50°W, where central pressure read 29.68 inches (100.5 kPa). Over central Canada and the northern Plains of the United States spread another frontal system whose converging course with the *R-34* would cause much trouble three days hence. At the moment, little heed was paid to it; the depression south of Newfoundland was to occupy all attention.

The first day of the trip experienced a relatively smooth passage through the high-pressure area, though extensive cloud strata and sea fog prevailed as a result of the northerly flow of unstable air from the arctic region. Only occasional sightings of the sea were possible. About noon of the third, with the ridge line of the anticyclone passed, the skies cleared, giving visibility of some 40–50 miles. At 1:30 P.M., the log reported: "Conditions now very pleasant. Blue skies above and blue sea below — nice and warm."

On the 1:30 P.M. weather map on July 3, the bow-shaped occluded front of the depression in the western Atlantic had advanced northeast, reaching to latitude 49.5°N. In the early afternoon the *R-34* began to sense the approach of the low-pressure system; cirrus clouds appeared and thickened, and the wind quickened from the southeast, raising "white horses" on the surface of the sea. At 2:10 P.M., "wind rising — sea beginning to get rough — visibility one mile," according to the log. Wind speeds were estimated at 30 knots (56 km/h), and by 4:30 P.M. they had increased to an estimated 45 knots (84 km/h) and were "backing" from southeast to east. At this time the *R-34* was close to 40°W, or directly on the longitude of the southern tip of Greenland, though 450 miles distant. By 7:00 P.M. they were in the thick of the approaching storm area, "all visibility gone, and thick fog everywhere. It is unpleasantly cold." Ten minutes later the ship was struck by "a fierce squall. Heavy rain." The log continued: "The rain is driving through the roof of the fore car in many places, and there is a thin film of water over chart table. The wind is roaring to such an extent that we have to shout to make ourselves heard."

This was the frontal system of the first depression. A secondary had formed on the cold front extending to the southwest, but this low pressure was too far south to affect the *R-34* seriously. By steering a northerly course, skirting the first depression, and taking advantage of the easterly flow in that quadrant, the skipper avoided what might have been a worse weather experience. The moon appeared around 9:00 P.M. as the ship left the zone of clouds and rain. Then a more southwesterly course was set to strike Newfoundland in the vicinity of St. John's.

The *R-34* encountered the usual fog when closing on the coast of Newfoundland, which was sighted at 12:50 P.M. on the fourth. A fresh tail wind soon picked up and pushed the ship along.

The attention of the meteorologists now turned west to the North American frontal system, which had moved eastward to the Labrador-Quebec border. Its warm front extended southeast across the Maritime Provinces with some clouds in advance, but no rain. The cold front bent westward from central Quebec to the Upper Great Lakes area. This was a vigorous front with an accelerating southeast rush.

The *R-34* made the leg from Newfoundland to central Nova Scotia without weather incident during the night and morning of the fifth, through the warm front of the approaching weather system. The Bay of Fundy appeared shortly after noon. At this time the cold front to the northwest, having crossed the St. Lawrence Valley, extended its influence over central New Brunswick and northern Maine. The front and the *R-34* were on a collision course. Things began to happen soon, according to the log:

1.30 P.M. — We begin to notice distinct evidence of electrical disturbances.

2 P.M. — Though the sky has not got much worse, atmospherics have become very bad. Severe thunderstorm can be seen over New Brunswick, moving south down coast. This storm looks very large, and appears to be moving rapidly. Scott turns left-handed off his course towards Nova Scotia to avoid it, but storm also extends eastward. He puts on all engines to try and get away from it, and orders are given to stow away all loose valuables.

2.45 P.M. — Caught in violent squall on extreme outskirts of the storm. Ship very badly thrown about, rising 700 feet in one bump. We see these bumps or squalls passing over the water. They appear to be a line of circular squalls or "whirls" moving in line abreast at a very high speed. Storm almost tropical in its violence. Our first warning was when the helmsman pointed to the compass card, which was spinning round like a top.

Harris thinks that had we been caught in the centre of this storm, the bumps would have been so severe that the ship might have been damaged in the air.

* * *

4 P.M. — Feeling very tired, so turn in and sleep soundly, despite the creaking of the girders in the keel, and loud buffeting of the wind against the outer cover.

6 P.M. — Dodging another colossal thunderstorm. Had to haul in W.T. aerials, as attempts to use wireless resulted in two-inch sparks owing to highly charged atmosphere.

Ship again badly thrown about — very unusual temperature bumps. Come to the conclusion that a comfortable hammock is indeed the best place on occasions like these.

* * *

Particularly fine electrical disturbance sunset.

* * *

9.20 P.M. — Durrant again lowers his aerials, but they charge up too quickly, and are at once reeled in.

9.30 P.M. — Violent temperature "bumps," evidently caused by rapid variation of sea temperature beneath us. Ship is first lifted 400 feet and then dropped 500 feet — measured on our aneroid. Scott, who has his head out of a window in the forward car, states that he saw the tail of the ship bend under the strain, whilst her angle is so steep at one moment that Cooke, resting in his hammock in the keel, is unable to get out for a minute or two, as he is head downwards at the time. Standing more out to sea, and running on all five engines to get further from this locality, a heart-breaking manoeuvre, as it will reduce still further our depleted petrol supply. Harris has experienced something of this kind before, when aboard the S.S. *Montcalm* in these regions. The sky was completely clear when the storm broke, wind practically calm, sea glassy, and moon brilliant. For a few seconds the warm air, which seemed pine scented (although we were well out to sea), was suddenly succeeded by very cold air, and it is these rapidly rising warm currents which throw the ship about with such violence.

Thunderstorms by day are bad enough, but at night they are particularly unpleasant, and the ship vibrates from bow to stern. We wear our parachutes, and life-belts are all ready.

Our only bottle of brandy fell out of the chart locker with a crash during one of these vertical bumps — fortunately without breaking.

In view of the buffeting by the winds and the diminishing fuel supply, the commander of the *R-34* first asked Halifax in the morning and Washington in the afternoon to dispatch a destroyer to tow the *R-34* if it needed to ration its fuel supply. The reply from Halifax was negative, because no ship was available, but the U.S. Navy ordered two destroyers to the waters south of Cape Cod. The *R-34* headed for Chatham, at the tip of Cape Cod, with the possibility of landing there or at Montauk Point at the eastern end of Long Island. The weather improved after midnight, as it usually does in a thunderstorm situation, and the remainder of the night and morning brought no more weather incidents.

Chatham, the first United States soil, was sighted at 4:10 A.M., Martha's Vineyard at 5:30 A.M., and Montauk Point at 7:20 A.M. Two hours later the *R-34* arrived over Hazelhurst Field (soon to be renamed Mitchel Field), and the ship touched down at 9:54 A.M.

The journey of approximately 3260 miles had consumed 108 hours, 12 minutes, or four and a half days, at an average of about

30 mi/h. The log closes: "We have 140 gallons of petrol, or two hours at full speed, so we couldn't have cut it much finer, and are lucky to get through."

The return flight from Long Island to Pulham, Norfolk, England, consumed 75 hours and 3 minutes, or 3 days, 3 hours, and 3 minutes, about 70 percent of the time required for the westward voyage. The return speed at an average of 43.4 miles per hour reflected the push of westerly winds. No major weather incidents occurred on the return trip. Only one frontal system was charted over the North Atlantic on July 10–12, and this weakened considerably while progressing eastward.

First Liquid-Propelled Rocket Flight — March 16, 1926

The Ward family farm on Pakachoag Hill in Auburn, Massachusetts, was the site of Robert H. Goddard's first successful test of his rocket on March 16, 1926. The site of this epic event is now marked by a granite monument placed by the American Rocket Society to commemorate the progenitor of modern jet aircraft and intercontinental missiles. Goddard, a professor of physics at Clark University in nearby Worcester, told of the attending circumstances in his diary:

> March 16. Went to Auburn with Mr. Sachs in morning. Esther and Mr. Roope came out at 1 P.M. Tried rocket at 2:30. It rose 41 ft. and went 184 ft. in 2.5 sec. after the lower half of nozzle had burned off. Brought materials to lab. Read *Mechanics, Physics of Air*, and wrote up experiment in evening.
>
> March 17. The first flight with a rocket using liquid propellants was made yesterday at Aunt Effie's farm in Auburn.
>
> The day was clear and comparatively quiet. The anemometer on the Physics lab was turning leisurely when Mr. Sachs and I left in the morning, and was turning as leisurely when we returned at 5:30 P.M.
>
> Even though the release was pulled, the rocket did not rise at first, but the flame came out, and there was a steady roar. After a number of seconds it rose, slowly until it cleared the frame, and then at express-train speed, curving over to the left, and striking the ice and snow, still going at a rapid rate.

It looked almost magical as it rose, without any appreciable greater noise or flame, as if it said, "I've been here long enough; I think I'll be going somewhere else, if you don't mind." Esther said that it looked like a fairy or aesthetic dancer, as it started off. The sky was clear, for the most part, with large shadowy clouds, but late in the afternoon there was a large pink cloud in the west, over which the sun shone. Some of the surprising things were the absence of smoke, the lack of very loud roar, and the smallness of the flame.

The cooperative observer of the U.S. Weather Bureau at Worcester registered a minimum of 16°F (−9°C) in the morning and an afternoon maximum of 40°F (4°C). A trace of precipitation was recorded during the 24 hours, time not specified.

First Solo Airplane Flight Across the Atlantic: The <u>Spirit of St. Louis</u> — 1927

This is Thursday, the 19th of May, exactly one week since I landed in New York. The sky is overcast. A light rain is falling. Dense fog shrouds the coasts of Nova Scotia and Newfoundland, and a storm area is developing west of France. It may be days, it may be — I feel depressed at the thought — another week or two before I can take off. I wouldn't be so concerned about the weather if the moon weren't already past full. Soon it won't be any use to me.

With these words, Charles A. Lindbergh, the one-time stunt pilot, barnstormer, army flier, and airmail pilot, expressed his exasperation with the delays caused by the adverse weather and his impatience to be off on his contemplated flight to Paris in hope of winning the $25,000 prize that had been offered by a New York hotel owner for the first to make a nonstop flight from New York.

For over a week slow-moving frontal systems had blanketed the Northeast and the Atlantic Provinces of Canada with fog and low clouds, and now a stationary front lay right over the New York area, running from a low-pressure center over the Province of New Brunswick to another sprawled across Pennsylvania and the Ohio Valley. In view of the unfavorable conditions in prospect, Lindbergh, along with several companions, decided to spend the day by paying a visit to the Wright aircraft factory in New

Jersey and then take in a performance of the musical *Rio Rita* on Broadway in the evening.

When back in New York City, Lindbergh made a phone call to James "Doc" Kimball, the chief forecaster at the New York Weather Bureau and a leading authority on North Atlantic weather. Kimball reported that local conditions would be improving in the morning, the fog along the coast should clear as the low-pressure center near Newfoundland receded to the northeast; and in its wake a high-pressure cell north of Bermuda would expand north into the transatlantic steamship lanes.

The decision was quickly made to skip the evening entertainment and head for the Long Island site where the *Spirit of St. Louis* was sheltered, at Curtiss Field in Mineola. Soon after their arrival at the hangar, the rain stopped, so the plane was brought out. It was necessary to haul it behind a truck to the adjoining Roosevelt Field, where the runway was sufficiently long for the takeoff of a heavily loaded aircraft. After a fitful attempt to grab a little sleep nearby, Lindbergh arrived at the field a little before 3:00 A.M. on May 20, 1927: "Clouds are low. It's hazy, and light rain is falling," he later wrote in his book of reminiscences, *The Spirit of St. Louis*, a reconstruction of the flight hour by hour.

The weather conditions prevailing while the plane was being readied were described in further passages:

> I slip out through the big, half-open door, and stare at the glowing mist above Garden City. That means a low ceiling and poor visibility — street lights thrown back and forth between wet earth and cloud. The ground is muddy and soft. Conditions certainly aren't what one would choose for the start of a record-breaking flight. But the message from Dr. Kimball says that the fog is lifting at most reporting stations between New York and Newfoundland. A high-pressure area is moving in over the entire North Atlantic. The only storms listed are local ones, along the coast of Europe.
>
> ★   ★   ★
>
> The wind changed at daybreak, changed after the *Spirit of St. Louis* was in take-off position on the west side of the field, changed after all those barrels of gasoline were filtered into the tanks, changed from head to tail — five miles an hour tail!
> A stronger wind would force me to the other end of the runway.

But this is only a breath; barely enough to lift a handkerchief held in the hand. It's blowing no faster than a man can walk. And if we move the plane, it may shift again as quickly as it did before. Taking off from west to east with a tail wind is dangerous enough — there are only telephone wires and a road at the far end of the field — but to go from east to west would mean flying right over the hangars and blocks of houses beyond — not a chance to live if anything went wrong. A missing cylinder and —— "Hit a house. Crashed. Burned." —— I can hear the pilots saying it — the end of another transatlantic flight.

The *Spirit of St. Louis* roared down almost the entire length of the mile-long runway before lifting its 5000-pound burden of plane and gasoline just in time to clear the wires at the end of the field by 20 feet. It was 7:52 A.M. and the course was set for Paris, some 3000 miles distant.

Weather conditions at takeoff time can be checked directly across the road from Roosevelt Field at the then-military airbase, Mitchel Field, from an observation taken at 7:40 A.M. The sky was overcast with stratocumulus clouds moving from the south. Light fog prevailed, but actual visibility in miles was not yet a recorded element in a weather observation. Wind movement came from the south at 4 mi/h (6 km/h), the barometer showed a very slow rise. The temperature stood at 55°F (13°C), and the relative humidity was high at 86 percent. This observation indicated that the clearing front had not yet reached the Mitchel area and conditions remained much as they had been for the past twenty-four hours.

But back on Manhattan Island in New York City, 23 miles to the west, weather conditions were improving minute by minute. At 8:00 A.M. the wind freshened out of the northwest at 15 mi/h (24 km/h). The temperature on the rise read 53°F (12°C). The barometer stood at 29.96 inches (101.5 kPa) and was rising, while the sky had broken to only partly cloudy. The cold front bringing the improving conditions moved through New York City while the *Spirit of St. Louis* was being readied on the field. The increasing westerly circulation would not only clear the skies and dispel the fog over Long Island, but give a quartering tail wind component to push the plane along the first leg of the trip.

The morning weather map for May 20, 1927, based on reports for an hour after Lindbergh's takeoff, showed the situation evolving as Dr. Kimball had forecast. The entire frontal system causing the unfavorable weather along the Atlantic Coast moved across Newfoundland in the north and off the New England coast in the south. Slow clearing set in from New York to Nova Scotia. The occluded front east of Newfoundland ran from a center south of Greenland between latitudes 45°W and 50°W well east of Cape Race, Newfoundland, and Cape Sable, Nova Scotia, the two promontories of the region. The Bermuda–Azores high-pressure area, now north of a normal position with its center about 45°N, extended an arm northeast toward the British Isles.

Lindbergh would have to pass through the occluded front in mid-Atlantic, a disturbed area, and then gain the favorable conditions expected in the expanding high-pressure system. Dr. Kimball had access only to reports from the steamers plying the North Atlantic lanes, and Lindbergh's planned "great-circle" route would take him several hundred miles north of the steamship lanes, into an area without radio weather reports.

After Lindbergh crossed the Thames River in eastern Connecticut, conditions began to improve. The morning mists dissipated, increasing the visibility, and the clouds commenced to rise and break, giving a higher ceiling. When he was over Cape Cod after two hours' flying, the sun burst through the overhead cockpit window and he was "flying under a dazzling blue field of sky surrounded by gray-bottomed clouds and hemmed in by glacial mountains of blinding white." The good flying conditions held through the fifth hour, when the *Spirit of St. Louis* was passing over Nova Scotia. Then clouds began to gather ahead and bumpy conditions were encountered. Some light rain squalls and turbulence prevailed during the seventh hour, over northern Nova Scotia, probably caused by the unstable polar air pouring over the uplands of the Gaspé Peninsula and northern New Brunswick. All this bumpiness subsided before Lindbergh left the mainland of North America. The air became crystal clear over the coast of the Cape Breton Islands, with only a few high cirrus clouds appearing during the water stretch over the Gulf of St. Lawrence to Newfoundland.

The *Spirit of St. Louis* roared over St. John's, the capital of Newfoundland in the southeast corner of the island, in the hazy light of sunset, with the ceiling unlimited and the visibility at 10 miles. The small plane had covered eleven hundred miles in eleven hours — one third of the flight. All was going well, with a favoring tail wind of about 30 mi/h (48 km/h), as Lindbergh bade farewell to North America and headed on the great circle route over the broad Atlantic.

Soon icebergs appeared in the Labrador Current below, with accompanying fog. "At first it doesn't hide the denser whiteness of the icebergs, but makes their forms more ghostlike down below. Then, the top of the veil slopes upward toward the east — a real fog, thick, hiding the ocean, hiding the icebergs, hiding even the lights of ships if there are any there to shine," he wrote. A ship was along his line of flight at approximately this time; it reported a temperature and dew point of 40°F (4°C), indicative of fog conditions.

During the fourteenth hour, the height and size of the clouds increased until they resembled huge mountains of condensed vapor. When he was flying through their swelling tops and updrafts, the air became quite rough. It grew colder and then ice began to form on the struts and the leading air-foil edges. During the fifteenth hour, this threat lessened as the coating of ice became thinner, began to evaporate, and finally disappeared. Lindbergh flew around the cloud mountains when he could or through valleys between them to avoid the icing conditions.

At the beginning of the sixteenth hour, the clouds broke in places and the moon peered through. Visibility and ceiling were now unlimited outside the clouds. Though in the next hour clouds appeared at all levels, cirrus, cumulus, and stratus, conditions improved: "The haze is almost gone . . . the clouds are no longer impassable barriers of ice. They're only opaque masses of air. I can fly through them if the compasses hold steady, drop down into them, keep right on heading eastward though they rise Himalaya-high."

Lindbergh had passed from the influence of the Labrador Current and into the atmosphere over the warmer waters of the Gulf Stream, he reasoned. All thought of turning back because of unfavorable weather was put away: "Now my anchor is in Europe; on a continent I've never seen. It's been shifted by the storm behind me, by the moon rising in the east, by the breaking sky and warmer air, and the possibility that the Gulf Stream may lie below. Now, I'll never think of turning back." The thick weather since leaving Newfoundland was caused by the principal storm system whose center had moved to a position near the tip of Greenland, with an active trough of low pressure trailing southward across Lindbergh's route.

At 5:52 A.M., GMT, the halfway mark was reached — eighteen hours out of New York, eighteen hundred miles accomplished and eighteen hundred to go, probably close to 35°W. With the coming of daylight, the cloud masses broke. Lindbergh was able to descend to get a look at the ocean and found that a strong northwest wind was blowing, of gale force, perhaps as much as 50 to 60 mi/h (80 to 97 km/h), he thought. But fog reappeared and blind flying was necessary for a while. "Will the fog never end? Does this storm cover the entire ocean? Except for that small, early morning plot of open sea, I've been in it or above it for nine hours. What happened to the high pressure area that was to give me a sunny sky?" he queried. The short stretch of clearing weather probably occurred when he had passed east of the trough of low pressure and had not yet reached the occluded front, now bowed well to the east, as this type of front often does, and lying close to 35°W at this time. The high-pressure area still lay ahead, but it was neither as strong nor as far north as Lindbergh had hoped.

During the twenty-second hour the thick fog dissolved and the sea appeared below. Alternate periods of scattered fog down to the waves and open patches in the cloud deck with blue sky above ensued. The ocean was not so wind-whipped and the flow shifted more to a full or quartering tail wind, west to northwest. "The wind continues to decrease. Angular rays from the sun spread through crevasses in clouds ahead. I climb to 500 feet. That keeps me above most areas of fog, and below most of the heavier cloud layers. Now and then I have to fly blind for a few minutes, but never for long."

Conditions continued to improve, with ocean and horizon

almost always in view. The *Spirit of St. Louis,* now well east of the front, enjoyed anticyclonic conditions afforded by the northeast extension of the Azores High. During the twenty-sixth hour, rising cumulus clouds dotted the sky with their puffs of white. Then light haze replaced midday's crystal clarity, and gray areas in the distance marked scattered rain squalls.

Fishing boats appeared during the twenty-seventh hour. "Patches of blue sky above me are shrinking in size. To the north, heavier storm clouds gather." Lindbergh was now on the eastern side of the rather narrow ridge of high pressure, putting him in a northerly flow of unstable polar air that is conducive to rain showers and squalls at this time of year. "I keep scanning the horizon through the breaks between squalls. Any one of those rain curtains may hide a ship or another fishing fleet. The air is cool, fresh, and pleasantly turbulent. I fly a hundred feet or so above the ocean — now under open sky, now with rain streaming over wings and struts."

At the end of sixteen hours' flying time from Newfoundland, a strip of low land appeared that Lindbergh soon recognized as Valentia and Dingle Bay, on the southwest coast of Ireland. His navigating proved excellent; he was only three miles off his planned great circle route, and he was two hours ahead of schedule, thanks to the favoring tail winds much of the way across the Atlantic. "The sun was still well up in the sky; the weather clearing" at his happy landfall.

With good visibility and land in sight for guidance, Lindbergh sped on the final leg of his epic trip. "The wind is strengthening, and tail. Cumulus clouds mottle the sea with their shadows. Scattered squalls, one after another, emerge from haze — light squalls, light haze, a clearing sky! I can see almost to the horizon." This came as a welcome change from conditions over the open Atlantic several hours back. England's Cornish coast now lay only two hours away. Surface winds over Ireland and southern England were northwest at 20 to 25 mi/h (32 to 40 km/h), and partly cloudy skies prevailed on the 1:00 P.M. weather map.

At this time, a barometric depression, centered over the North Sea between Scotland and southern Norway, caused the blustery winds over the British Isles. Since the cold front of this system had swept eastward into Germany and eastern France, the skies were clearing along the Channel coast and inland as if to welcome the lone aviator in the *Spirit of St. Louis* with favorable conditions:

> Only three hundred miles to Paris. The horizon is sharpening, and the sky ahead is clear.

*   *   *

> The sun almost touches the horizon as I look down on the city of Cherbourg. Ahead the sky is clear.

*   *   *

> I'm still flying at four thousand feet when I see it, that scarcely perceptible glow, as though the moon had rushed ahead of schedule. Paris is rising over the edge of the earth. It's almost thirty-three hours since my take-off on Long Island. As minutes pass, myriad pin points of light emerge, a patch of starlit earth under a starlit sky — the lamps of Paris — straight lines of lights, curving lines of lights, squares of lights, black spaces in between. Gradually avenues, parks, and buildings take outline form; and there, far below, a little offset from center, is a column of lights pointing upward, changing angles as I fly — the Eiffel Tower. I circle once above it, and turn northeastward toward Le Bourget.

*   *   *

> At one thousand feet I discover the wind sock, dimly lighted, on top of some building. It's bulged, but far from stiff. That means a gentle, constant wind, not over ten or fifteen miles an hour.

Lindbergh made a near-midnight landing at Le Bourget Airfield under ideal weather conditions: light winds, starlit sky, and excellent visibility. It was 5:21 P.M., New York time, and he was thirty-three and a half hours and 3300 miles away from his take-off on another continent.

The weather gods favored Lindbergh. He chose the month of May, when conditions are normally at their most tranquil over the northern portion of the North Atlantic. He had an experienced forecaster who perceived the chance of a favorable weather situation unfolding. He kept in touch with his weather advisor and played a hunch at the proper time, when his rivals were indecisive. During the first leg of the trip, though some turbulence was experienced over Nova Scotia, conditions continued mainly favorable with tail winds helping to preserve precious gasoline. He crossed the southeast tip of Newfoundland in twilight and with good visibility, enabling him to get a sure visual fix on St. John's at the departure point from land on the transatlantic hop.

Despite the weather hazards during the first half of the overwater route, when passing through the disturbed conditions in the trough of the low-pressure system, Lindbergh was able to keep to his projected course by accurate instrument navigation. No long period of crosswinds was met that might have pushed him off course, and no mountains or land hazards lay along the route to complicate a safe altitude level.

During the last half of the flight, the *Spirit of St. Louis* enjoyed anticyclonic conditions as predicted, with broken clouds and fair weather, still with a quartering tail wind. Lindbergh was able to make a visual sighting of the southwest Irish coast to determine his exact position. The flight across southern England with continued air-to-ground visibility led him to the French coast at dusk. Then from 50 miles away in a clear sky, the lights of Paris guided him to his rendezvous with history.

First Transatlantic Balloon Flight — 1978

Double Eagle

The challenge of a flight across the North Atlantic Ocean to Europe has long intrigued visionary balloonists with a penchant for distance flying. Since the first serious attempt in 1873, fourteen flights were made without achieving the goal. One established a new distance record for sustained flight of 2745 miles, and another was forced to ditch into the ocean as a result of developing a rip in the helium envelope when only 110 miles off the French coast.

Inspired by reading a February 1977 article that described a recent attempt to cross the Atlantic, Maxie Anderson of Albuquerque, New Mexico, began to think about the possibility of spanning the ocean in a specially designed helium balloon. He mentioned the idea to a close friend and fellow-balloonist, Ben Abruzzo, who eagerly agreed to participate. After several months of hectic preparation for a launch before the onset of the bad weather season, Anderson and Abruzzo took off in *Double Eagle* from Marshfield, Massachusetts, on September 9, 1977. A delay of several hours in the scheduled departure caused the balloonists to miss the favorable pressure patterns with helping winds that meteorologist Robert B. Rice of Weather Services International of Bedford, Massachusetts, had forecast and had urged them to take advantage of.

Instead of finding smooth sailing on an easterly course at relatively high altitudes, *Double Eagle* became involved in a rapidly developing, low-level cyclonic storm system between southern Greenland and Iceland. Heavy rain mixed with snow, below-freezing temperatures, and severe turbulence caused extreme stress to the balloon structure and acute discomfort for the crew.

The cyclonic wind pattern around the storm center carried *Double Eagle* in a counterclockwise, circular loop northward and westward instead of progressing eastward as planned. After a flight of 65 hours and 43 minutes, the balloonists were forced to ditch into the ocean at a point only three miles west of Iceland. A rescue team with helicopter, called to the scene as a precaution, pulled the exhausted men to safety from a small skiff that had been carried below the balloon gondola to serve as a lifeboat in such an emergency.

Flight Planning

The fundamental factor in Rice's flight planning was the utilization of an altitude not normally associated with long-distance ballooning, with 20,000 feet or more the goal. Four factors that prevail at high altitudes over the North Atlantic Ocean make this desirable: (1) Wind-flow patterns are much steadier and less subject to unanticipated changes at these levels. (2) Wind speeds generally increase with elevation, making for a faster flight. (3) Pressure systems usually slope westward, enabling the balloon to fly over bad weather at the surface. (4) The cirrus-type clouds prevailing at high altitudes have a less cooling effect on the gas envelope than do the more opaque clouds prevailing at lower levels.

Nevertheless, flying at such high altitudes requires more discipline and determination from the balloonists because the constant employment of oxygen is required and the physical discomforts from cold are more taxing.

The desired weather-map situation requires a north–south ridge of high pressure of some stability moving eastward off the North American continent. The vertical structure of such ridges usually slants backward from the surface toward the west for several hundred miles. A launch into the trailing lower portion of the ridge provides a wind direction that will carry the balloon to the northeast toward the higher latitudes necessary to obtain a route leading to Europe. It is imperative to travel far enough north to avoid the large sprawling Azores–Bermuda high-pressure area over the central Atlantic with its contrary air flow from the east. After a twenty-four-hour flight at relatively low levels, the balloon ascends through the slanting strata of the high-pressure ridge and obtains a position just ahead of the upper-level ridge, where northwest and west winds carry the balloon eastward toward its goal.

The high-pressure ridge takes the balloon only part way to Europe. A second feature is needed, in the form of a blocked or slow-moving low-pressure system located in the northeast corner of the Atlantic Ocean lying north of 55° to 60°N. Winds from a westerly direction found in the southwest quadrant of this system should speed the balloon onward in the direction of the British Isles and Europe.

From the weather standpoint, the initial flight through the high-pressure ridge provides fair weather with plenty of sunshine, while in the southwest quadrant of the low-pressure area clear or clearing conditions without local disturbances prevail.

Double Eagle II
Within two days of the rescue off Iceland, Max Anderson began making plans for a second flight. During a skiing weekend in February 1978, he suggested the idea to Ben Abruzzo and after some thought the latter agreed. When it was decided to take along a third man, Larry Newman, a friend of Abruzzo's, volunteered. Though he was an airplane pilot, he had no experience in bal-

looning, but his knowledge of navigation and communication would be helpful, and the presence of a third party would dispel some of the psychological stress between only two persons that existed on the first flight.

Double Eagle II was a larger version of the first balloon, having a capacity of 160,000 feet of helium, 59,000 more than its predecessor. When inflated, the envelope would be sixty-five feet in diameter and ninety-seven feet high. It was planned for a flight lasting seven days. Richard Schwoebel designed the balloon and acted as general manager for the entire project.

A launch site near Presque Isle in northeast Maine was selected to give a more northerly departure into favorable wind currents and save many hours of flight. A target takeoff date for mid-August was set.

As the date approached, forecaster Rice informed the balloonists on Tuesday, August 8, that conditions appeared to be developing for Friday that would provide a pattern consistent with all preliminary conditions specified.

SATURDAY, AUGUST 12, 1978 — The launch of *Double Eagle II* took place on Friday at 8:43 P.M. (EDT) or 12:43 A.M. (GMT) on Saturday. This was three hours later than planned and nearly caused a failure to fulfill the flight pattern, which called for the balloon to be launched forward of the ridge line of an eastward-moving high-pressure area. Though the ridge at the surface had passed beyond the longitude of Presque Isle, the balloon rose during the night and achieved a position slightly ahead of the ridge line at 5000 feet. Light westerly winds in the heart of the ridge carried the balloon eastward as desired. Over the Gulf of St. Lawrence, *Double Eagle II* showed a tendency to sink; it was with some difficulty that the flight level was maintained.

SUNDAY, AUGUST 13 — During the morning the balloon skirted the south coast of Newfoundland at an altitude of about 10,000 feet. It appeared to have fallen slightly behind the ridge line, since winds were south of west and still light. This carried the balloon along with a gradual turn to the northeast. In the thirty-eighth hour of flight, at 3:00 P.M., the balloon passed over St. John's in the southeast corner of the island and headed over the open Atlantic. A low-pressure vortex near Iceland, ahead of the high-pressure ridge, was behaving nicely and moving steadily toward the desired position north of the British Isles.

MONDAY, AUGUST 14 — The path of the balloon continued northeast, holding a position behind the ridge line by perhaps 50 nautical miles. A flight altitude between 13,500 and 16,000 feet was maintained. The three-hour delay in launching had cost some 60 to 80 miles of travel.

On the noon weather map an unanticipated development became apparent and threatened to disrupt the flight plan. The predicted development of a low-pressure trough following the high-pressure area took shape over Labrador and Newfoundland, but in addition a wave or secondary disturbance formed on the old polar front southeast of Nova Scotia and soon took control of the regional circulation. This diminished the separation of the balloon from the new storm center by about 12 hours, much less than was desired.

TUESDAY, AUGUST 15 — When the balloon was little more than halfway between Newfoundland and Ireland, a very real concern arose regarding the rapid development of the storm system to the rear. *Double Eagle II* climbed all day and by evening once again attained the high-pressure ridge line at about 17,500 feet. The lower circulation far below the balloon was already taking on a menacing cyclonic curvature. Had the balloon been down there, it would have turned north and in all probability would have been absorbed into the storm circulation. "The afternoon of the 15th was one of the longest in the memory of this meteorologist," Bob Rice recalled. By late evening, however, satellite pictures made it clear that the balloon had attained the ridge at about 25°W and was passing above it as the flight "commenced a glorious turn to the east." This was a "clear victory for the pre-flight high altitude profile decision," declared Rice. Now *Double Eagle II* was safely accelerating eastward, while the developing massive ocean storm was left behind. Ireland lay about 700 miles away at midnight.

* * *

WEDNESDAY, AUGUST 16 — Though the balloon was now in a fair-weather area west of Ireland, an unexpected problem soon arose. A sudden loss of altitude carried the balloon down about 20,000 feet, or a descent of four miles. This presented a novel situation about which meteorologists later hypothesized that the balloon drifted into a column of subsiding air of warmer temperature than the equilibrium-level cooler-air stratum in which the balloon had been traveling. The incident "prompts the thought that the cause of failure of trans-Atlantic balloon flights does not necessarily have to be 'bad' weather," Rice later philosophized.

After stabilizing at the lower level and coming under the influence of the afternoon sun, *Double Eagle II* climbed back to an altitude in excess of 24,000 feet by midevening as it was nearing the coast of Ireland. Thereafter, it became mainly a question of the durability of the balloon, for the weather ahead proved favorable. Landfall on the Irish coast was made over Clare Island at about 9:00 P.M. (GMT) and they crossed Ireland during the night.

THURSDAY, AUGUST 17 — Early on the final day, *Double Eagle II* left the high-pressure ridge well behind and came under the steering influence of the low-pressure system ahead, which had progressed from Iceland to a position off the northern coast of Scotland. In its southwest quarter winds blew from a favorable direction, mainly northwesterly and westerly; these carried the balloon southeast across Wales, southern England, and the English Channel to the control of another high-pressure, fair-weather system stalled over western Europe. The balloon maintained an altitude of 12,000 to 15,000 feet while crossing the Channel from Bournemouth to Le Havre and then descended over the French countryside to take advantage of the lower-level winds. A favorable arcing toward the east followed in the gentle wind flow prevailing while the sun was sinking in the west.

Double Eagle II landed safely in France at Miserey near Evreux, about 60 miles west of Paris, at 5:49 P.M. local time. Weather conditions were fair with light and variable winds associated with the stalled high-pressure area now centered to the southwest on the Brittany coast. The total flight time was 137 hours, 5 minutes, and 30 seconds.

Airship Losses

Loss of the U.S.S. Shenandoah: Squall Line — 1925

The name of the lovely valley in western Virginia, derived from the Indian phrase meaning "Daughter of the Stars," was given to the first rigid, lighter-than-air ship to be built in the United States. The U.S.S. *Shenandoah* was a copy of German Zeppelin *L-49*, with a gas capacity of about 2,100,000 cubic feet. Commissioned on September 4, 1923, she underwent many successful flights, including a transcontinental trip.

Almost two years to the day after its first flight, the *Shenandoah* departed its base at Lakehurst, New Jersey, on the afternoon of September 2, 1925, for a goodwill mission to various state fairs in the Midwest. Some criticism was voiced in the press about sending the pride of the navy into an area whose atmosphere was prone to turbulence at this season of the year, but the will of the public relations people prevailed.

At the time of departure the weather map did not appear to hold any threat. A frontal system had passed eastward off the coast, and a wide high-pressure ridge covered the Appalachians, Lower Lakes, and Ohio Valley. One cell of high pressure lay over Ontario and Quebec and another over West Virginia. The culprit in the situation at this time was far to the west, in the form of an occluded weather front extending from northern Manitoba, south into Minnesota near International Falls, and then southwest, passing near Yankton, South Dakota, and Cheyenne, Wyoming. Over the eastern section of the proposed route, skies were clear to partly cloudy and winds were light.

At midnight the *Shenandoah* was cruising over the Allegheny Mountains of western Pennsylvania. The sky was mostly covered by broken clouds, but the air was calm. It remained warm at this hour because the ship had passed into the southerly circulation on the backside of the retreating anticyclone. The occluded frontal system still lay well to the west, about 400 miles distant, over Wisconsin and Minnesota. Lieutenant Joseph B. Anderson, the ship's aerologist, studied the midnight weather re-

The wreck of the Shenandoah Airship near Ava, Ohio, September 3, 1925. Official U.S. Air Force photo.

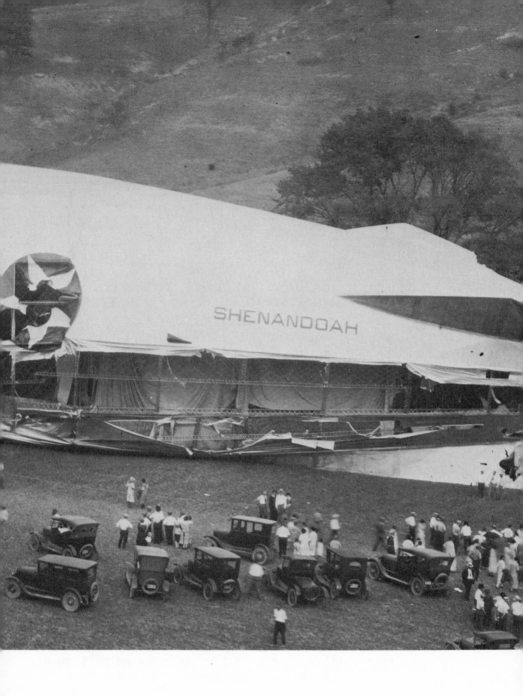

ports and handed his summary to Lieutenant Commander Zachary Landsdowne, the chief officer. No indications of imminent deteriorating conditions appeared along the projected route. An area of thunderstorms was spotted over the Great Lakes, but this was not expected to extend far southward. So Anderson turned in for some much-needed rest after a long arduous day in preparation and flight. He left word not to be called "unless something unusual comes up."

As the early morning hours of September 3 passed, the *Shenandoah* began to encounter bumpy conditions of increasing severity as it crossed into southeast Ohio close to the east-west path of U.S. Route 40. The occluded front had advanced to a line running from near Sault Sainte Marie to Milwaukee to Omaha, with an increasingly strong flow of warm tropical air prevailing over the Ohio Valley and Lower Lakes region. Coming from the southwest, the winds aloft greatly slowed the forward progress of the ship and eventually reduced her ground speed to near zero.

Lieutenant Commander C. E. Rosendahl, the navigator and third in command, kept a close watch on the weather:

> At the windows, Captain Lansdowne and several of the officers were watching the weather. We were then over Byesville, Ohio, making very little ground speed, with all five engines running, and had just decreased our altitude to 2100 feet by altimeter. To the northward and eastward a severe electrical display and heavy clouds stood out vividly several miles distant. I was told we had changed course during the mid-watch to avoid a thunderstorm. Other portions of the sky were now clear. Often on previous occasions the ship had been in strong winds that had materially reduced her speed over the ground, but now even with three of the engines speeded up and the other two at cruising speed, we made no gain and our drift increased. Captain Lansdowne was a thorough student of flying weather and although we had passed all evident danger, he remained on the bridge. The weather map on our departure had been considered a reasonably safe one as the disturbance centered near the Great Lakes should normally move on northeastward out of our path, and there were no indications of serious trouble along our course.
>
> The dull moon ahead was setting with a weakening glow. During my drift observations out the window, there appeared on

the starboard bow a thin dark streaky cloud just apparent in the dull moonlight. Lieutenant Anderson, the aerologist, seated at another window saw it at about the same time. Almost immediately the Captain came over to view this formation that was either coming towards us or building up rapidly. About this same instant, Chief Rigger Allen at the elevators reported the ship rising. He was told to check her. This he could not do. "She's rising at two meters per second and I can't check her, sir," he reported. The engines were speeded up. The inclination of the ship driving downward against the rising air current was considerable. With my flashlight I read it several times at 18 degrees. "Don't exceed that angle," ordered the Captain. "We don't want to go into a stall." Meanwhile at the beginning of the rise, the Captain had ordered the course changed somewhat to the left, i.e., to the southward. We then realized that we had run into a forming line squall — a line marked by the streaky roll cloud where winds of different temperatures and from different directions clash sharply and set up turmoil and vertical air currents both up and down. But even at this stage we felt no immediate concern for the safety of the ship as airships had plowed through storms before; though of course no one particularly relished being in any storm.

The time was now 4:20 A.M. The ascent began at an altitude of 1600 feet and continued at an average rate of 225 feet per minute. The ship leveled off at 3000 feet, rolling and pitching, and then rose even faster to a height of 6300 feet, the last part of the ascent being at the rapid rate of 1000 feet per minute. This altitude was well above the ship's pressure height of 4000 feet and the gasbags were in danger of exploding from the reduced pressure. Much helium was then valved, causing the *Shenandoah* to sink rapidly to about 3000 feet, where she was halted abruptly by another ascending column of air. A final and fatal rise began at 4:50 A.M.

The unequal force of the rising shafts of air carried the *Shenandoah* up to 3600 feet, the nose climbing faster until it reached an inclination of 30 degrees. The elevated position of the forward section was opposed to the inertial resistance of the remainder of the ship. The point of greatest stress in the rigid structure centered in the vicinity of frame 130. "With a terrific crashing of metal and a combination of noises hard to describe," Rosendahl recalled, the hull broke near frame 125, about 220 feet from the bow of the 660-foot-long ship.

At this very moment Aerologist Anderson reached the top of the ladder leading from the keel to the control car. He was horrified to see the braces and girders holding the car to the frame loosening and bending. For a while the two sections of the airship were held together by the control cables running along the keel, but soon they wrenched asunder. At the same time the control car, containing eight men, broke from the frame and plummeted to earth, killing all. Anderson witnessed all this as he struggled to gain a secure position on the catwalk of the ship. Meanwhile, the gasbag and frame of the forward section, relieved of the weight of the control car, rose high above the remainder of the *Shenandoah*. The forward gasbags had not ruptured, and seven men, including Anderson, took off on an aerial voyage. They zoomed up to 6000 feet before Rosendahl and his men were able to valve gas and gain control of their now free balloon.

Back on the main section of the ship another tearing and ripping of girders occurred near frame 100, almost amidship of the original hull. Now the *Shenandoah* was in three parts. The center section, weighed down by the burden of engine gondolas 4 and 5, dropped rapidly, "like an elevator with no brakes," and smashed into the side of a small hill. Four men in the hull part were injured but survived; four mechanics in the engine gondolas were killed.

The tail section, about 350 feet in length and weighed down by its three engines, also fell rapidly and crashed against a wooded hillside, where the engines were scraped off by the trees. The gasbag shot upward again, but soon settled into a small valley. The eighteen occupants scrambled out of the way of the collapsing frame and escaped.

The seven men in the balloon gasbag were buffeted by prevailing wind squalls. Plans were made to valve more helium and land, but winds in the vicinity of 25 mi/h (40 km/h) made this a precarious operation. After about an hour the wind died down sufficiently to permit a successful landing. At 6:45 A.M., Rosendahl, Anderson, and companions came down at Alva, Noble County, about 12 miles from Byesville, Guernsey County, where the control car landed.

Squall lines of the type that doomed the *Shenandoah* develop well in advance of an east-moving cold front. They are not apparent on a weather map until actual thunderstorms begin to flash and thunder. Their development is often rapid when a cold northwest current overruns a warm southwest current. Severe turbulence results with strong ascending and descending shafts of air interspersed in close proximity within the thunderstorm cell. It would appear that one rising current caught the nose of the *Shenandoah* in its upward thrust and subjected the rigid hull to unequal forces, resulting in its fracture. This type of atmospheric behavior has long been recognized as the major hazard for aircraft operating over the Midwest.

Loss of U.S.S. Akron: Thunderstorms and Turbulence — 1933

The decade of the 1930s was the heyday of lighter-than-air craft. The Zeppelin works in Germany were busy turning out famed craft that flew around the world and to South and North America. The United States entered the competition by building two large ships of 6,500,000 cubic feet capacity and 785 feet length. U.S.S. *Akron*, commissioned in 1931, had 1700 hours in the air to its credit by 1933. When it took off from Lakehurst Naval Air Station in central New Jersey at 7:28 P.M. on April 3, 1933, the wind at ground level moved lightly from the northeast at about 6 knots (11 km/h). A ceiling of about 300 feet prevailed, because fog had recently developed over the field. No bad weather was indicated locally, aside from the fog, nor had the weather predictions by Navy aerologists suggested unusually severe conditions that might endanger an airship flight. The *Akron* was off on a three- or four-day training mission, with Newport, Rhode Island, an early port of call.

The weather map at 8:00 A.M. (EST) on April 3, placed an occluded front bowing eastward from Lake Superior over Lake Ontario, through central Pennsylvania to a point of occlusion over central Virginia. The warm front of this system extended to near Cape Hatteras and then seaward. Another front dropped southward from the main low-pressure center over Wisconsin through Michigan, Illinois, and Missouri. Light rain showers were falling over the Appalachians along the eastern front. Winds over New Jersey were northeast at moderate speeds, bringing moist airstreams over the land.

Commander Frank C. McCord called a weather conference with his forecasters at 4:00 P.M. and the decision was made to take off, since the only unfavorable weather factor seemed to be the low ceiling over the field. At that very time meteorological events were in progress that were to produce a series of violent thunderstorms in a few hours over New Jersey and the adjacent ocean. As frequently happens, a secondary low-pressure center formed at the peak of the occlusion in Virginia sometime shortly after noon. With a supply of moist, tropical air to the east and cold air to the west, the new center developed energy rapidly and soon became the main low-pressure center. It moved northeast over Chesapeake Bay about sundown and headed for the south Jersey coast. Weather Bureau forecasters in Washington had not anticipated this development. Gordon R. Dunn, the chief forecaster at the nation's capital, testified that the first he knew of the storm's existence was the lightning he viewed from his home that evening.

The *Akron* first headed west, away from the fog area, and then south. At 8:30 P.M., when the airship was cruising over Wilmington, Delaware, sharp flashes of lightning were seen to the south at a distance of about 25 miles, and soon they were noticed to the west also. The *Akron*'s commander changed course, heading east and then northeast over New Jersey, away from the lightning areas.

Static on the radio prevented the complete reception of the 8:00 P.M. weather data, though enough was received to indicate a low-pressure center in the vicinity of Washington, D.C. This was the secondary storm center that formed during the noon hour over Virginia. As it developed northeast, thunderstorms of increasing severity broke out in advance as the warm air and cold air clashed ahead of the track of the storm center: at Washington by 7:35 P.M., at Baltimore by 8:05, at Philadelphia by 9:49, at Trenton by 9:55, and at Atlantic City, the station nearest the *Ak-*

U.S.S. Akron *over Union Station, Washington. Post Office and Government Printing Office, Washington, D.C.*

ron, by 10:14. The zone of instability was quite concentrated, no thunderstorm activity being reported westward at either Harrisburg or Reading.

U.S.S. *Akron* crossed the coastline, heading seaward in stable air, but by 11:00 P.M. intense lightning, both cloud-to-ground and cloud-to-cloud, enveloped the ship and severe turbulence was experienced, so the course was reversed to the west. Upon arrival at the coastline, in the vicinity of Barnegat Coast Guard Station, about midnight, the airship changed direction again, this time seaward to the southeast. The current wind at the surface was about 40 knots (46 mi/h or 74 km/h) from the northeast with rain, light fog, and heavy thunder occurring.

At 12:15 A.M., when off Long Beach Island on the central New Jersey shore, the *Akron* hit a descending current and lost altitude rapidly. Ballast was dropped, control regained, and the ship righted at a flying altitude of 1600 feet. About three minutes later she began to sink again, this time at the extreme rate of 14 feet per second when caught in another downdraft. Engines were given full power, causing the nose of the 785-feet-long ship to point upward at an angle of 20 to 25 degrees. Suddenly a severe shock was felt throughout the entire frame when the lower rudder and controls were torn away, apparently having struck the surface of the water while the ship was angling upward. The impact of the stern with the water resulted in the progressive destruction of the frame.

Soon, all of the *Akron* was down in the water, and the sea poured into the control gondola. The accident occurred close to 12:30 A.M., at a position about 19 nautical miles south-southeast of Barnegat Inlet. The wreckage floated for a while and when found later rested on the bottom of the ocean about 27 nautical miles southeast of Barnegat and due east of Little Egg Inlet. Investigators have speculated that, if the *Akron* had only had 100 feet more altitude at the time of her descent, the ocean impact of the stern might not have occurred.

Partly from lack of lifesaving equipment and partly from the coldness of the water, the loss of life was very heavy. Of the seventy-six persons aboard, only three were rescued from the water, these by a German freighter that had been close enough to see the lights of the *Akron* disappear into the sea.

The commander of the *Akron*, according to a reconstruction of the situation after a congressional inquiry into the disaster, was thought to be aware that a small low-pressure center was moving northeast from the Washington–Baltimore area. By heading out to sea originally, he might cross the projected path of the advancing storm well ahead of it and find more stable conditions over the ocean in the warm sector to the east. But when the weather became very turbulent around the ship and lightning played about, he assumed the center of the disturbance was close and reversed course to the west. Then, preferring to cruise over the sea than over the land under the turbulent conditions existing, he headed for the southeast, hoping to move into the rear of the storm. This apparently brought the ship on a collision path with the most turbulent sector of the storm center.

Loss of U.S.S. <u>Macon</u>: Turbulent Wind Gust — 1935

The *Akron*'s twin sister, *Macon*, was based at Sunnyvale in north-central California near San Jose. Her silhouette came to be well known to Pacific Coast residents while she logged 1798 hours in the air, mainly in the West.

After participating in naval exercises off the Southern California coast, U.S.S. *Macon* was returning to base on the late afternoon of February 12, 1935. While cruising at 2500 feet near Point

Sur Lighthouse, the airship began to encounter a deteriorating weather situation. Horizontal visibility remained good, however, for the lighthouse keepers had the airship under surveillance with binoculars at a distance of three miles.

The weather map on the morning of the twelfth placed the eastern cell of the Pacific High with pressure at 30.32 inches (102.7 kPa) at about 27.5°N and 140°W. A ridge extended northeast across northern California to a secondary high cell over southern Idaho at 29.98 inches (101.5 kPa). Wind flow from San Francisco Bay southward had a northerly component, and speeds were as high as force 6 (25 mi/h or 40 km/h), since a low-pressure center was near Yuma, Arizona, with pressure at 29.53 inches (100.0 kPa). Winds on the central California coast are normally at their peak in late afternoon, when the temperature differential between sea and land is greatest. There were no fronts in the vicinity, merely the usual northerly flow augmented by the existing pressure gradient.

The position and weather situation of the *Macon* in the late afternoon has been described with precision of wording by Richard K. Smith in his definitive study, *The Airships Akron and Macon*:

> Cruising at 2,500 feet, the *Macon* began to encounter a lowering ceiling and a series of rain squalls. Altitude was changed to 1,700 feet in an effort to duck under the weather. At 1630 a Luckenbach intercoastal freighter was observed below, paralleling the *Macon*'s course; a few minutes later the airship flew into a curtain of rain, and visibility dropped to less than a mile. Because of increasingly high fog off Cape San Martin, altitude was further reduced to 1,400 feet.
>
> * * *
>
> At 1656 Point Sur's lighthouse appeared in the gloom, flashing two points off the starboard bow. At this moment a shaft of sunlight suddenly pierced the weather in the west, and the watch in the control car could see seven cruisers on the surface, working against relatively heavy seas. At 1704 Point Sur was almost abeam. Wiley noticed that the *Macon*'s course was slowly bringing her closer to the coast, and preferring to keep her well out to sea and away from the rain-shrouded mountains, he ordered left rudder.

Coxswain William H. Clarke put his wheel over and was holding about five degrees rudder when what seemed like a sharp gust struck the *Macon*. The airship lurched violently to starboard; the helm was wrenched from Clarke's hands and went spinning wildly. In this same moment, Aviation Metalsmith First Class William M. Conover, who was at the *Macon*'s elevator controls, noticed the airship pass into a veil of haze, which alerted him for turbulent air. He had just recovered the airship from a shallow dive when he felt the gust. The control wheel was torn loose from his hands for a moment, but he quickly regained control.

*   *   *

At Point Sur, lighthouse keeper Thomas Henderson and his assistant, Harry Miller, were watching three cruisers when the *Macon* appeared out of the south. As she hurried along beneath the overcast, both men put their glasses on her. Just as the *Macon* came abeam of them, her stern took a sharp dip, and an instant later they were startled to see her upper fin literally disintegrate.

Henderson had his glasses trained on the airship's stern section; he was interested in the motion of her elevators. He saw the upper fin's leading edge lift away slightly from the *Macon*'s hull; then the fin suddenly, but progressively, flew into pieces.

They watched the *Macon* swing sharply away from them and fly off on a circular track to the south. As she made her turn, gray streams of gasoline and ballast trailed from her, and they heard dull explosions as her slip tanks struck the sea. Miller looked at his watch. The time was 1707. As the *Macon* came about on her southerly heading and was again almost broadside to them, Miller noticed that the rudder post and its rudder were sill sticking up above the airship's hull. A few seconds later the *Macon* disappeared into the overcast, but they were able to follow her track for awhile by her trail of splashing ballast and slip tanks. Then a curtain of rain cut off their view.

The forceful lateral impact of the blow caused the upper rear fin containing part of the rudder to tear loose from the ship's frame; as a result, shattered parts of the *Macon*'s structure punctured three gas cells in the stern section and helium filled the interior of the rear cone. Large quantities of fuel and ballast aft of amidships were dropped in an effort to trim the ship. This was the critical decision that sealed the fate of the *Macon*. She immediately began to shoot skyward — nose up, her stern heavy — past her critical pressure height of 2800 feet, where the automatic valves opened and released her helium into the atmosphere. The airship reached a peak altitude of 4850 feet within eight minutes of the initial accident and remained above her pressure height for sixteen full minutes, all the while pouring her precious life-gas into the air.

So much helium was lost in the ascent that the *Macon* was no longer buoyant. Now, there was no place to go but down, and down the airship sank, rapidly at first, at the alarming rate of 750 feet per minute, and then more slowly, at a rate of 150 feet per minute after more water ballast had been jettisoned. At 1700 feet the airship broke through the overcast. With the sea visible, word was passed to stand by to abandon ship. She hit the surface of the sea at 5:39 P.M., only 34 minutes after the wind gust had struck.

The weather and sea had become quite calm by the time the *Macon* disappeared under the water at 6:20 P.M. at a position about 13 miles northwest of Big Sur Light. In contrast to the *Akron* disaster, a good supply of lifesaving equipment was aboard, the water was relatively warm, and nearby cruisers rushed to the rescue. Only two of a crew of 83 were unaccounted for.

The weather-related disasters to the *Shenandoah, Akron,* and *Macon* did not end the navy's interest in developing this type of lighter-than-air ship. Some naval airmen remained enthusiasts and urged the construction of one more American ship. Congress authorized a small one, but the development funds were withdrawn when unreasonable delays developed in selecting a design. The outbreak of World War II ended the project while it was still on the preliminary drawing boards.

Meanwhile, the Germans were active with the *Hindenburg* and *Graf Zeppelin* and were building others. They seemed on the verge of proving the reliability of these commercial airships when disaster struck.

The Loss of the Hindenburg: No One Knows the Cause — 1937

Captain Max Pruss was a veteran airship man, having made ten trips to the United States and eight to South America in the pride of the Nazi regime, the zeppelin *Hindenburg*. Upon what was to be his last arrival in America, he described conditions on May 6, 1937:

> We had bad weather. About two o'clock we were over New York, made a few circles, and then went on to Lakehurst. Then we saw a big thunderstorm over New Jersey and knew we couldn't land and thought it better to go back to the sea. We went along the coast to Atlantic City and back, and we waited for the storm to blow over to the ocean.

The weather map on the morning of May 6 located a low-pressure center over eastern Lake Ontario with a cold front running southward through east-central Pennsylvania, then bowing to the southwest over central Virginia and southward to about Wilmington, North Carolina. The thunderstorms in the New Jersey area were caused by the passage of a line of instability accompanying the cold front late on the afternoon of the sixth. The frontal zone had apparently cleared Lakehurst by 6:00 P.M., when winds shifted into the west-northwest and gusted to 23 mi/h (37 km/h).

On the ground at Lakehurst, the weather situation was watched carefully by all concerned with landing the *Hindenburg*. The following exchanges took place between the two parties:

At 5:12 P.M. Naval Air Lakehurst radioed: THUNDERSTORM MOVING FROM WEST OVER STATION SURFACE TEMPERATURE 70 FALLING SURFACE WIND WEST SIXTEEN KNOTS GUSTS TWENTY ONE WIND SHIFTED FROM NORTH AT 1600 EST PRESSURE TWENTY NINE SIXTY FOUR RISING.

At 5:35 P.M. Captain Pruss queried: HOW VISIBILITY FROM LAKEHURST WESTWARD. The Naval Air Station replied: VISIBILITY WESTWARD EIGHT MILES UNSETTLED RECOMMEND DELAY LANDING UNTIL FURTHER WORD FROM STATION. ADVISE YOUR DECISION.

Captain Pruss answered: WE WILL WAIT REPORT THAT LANDING CONDITIONS ARE BETTER.

At 6:12 P.M. Lakehurst radioed: CONDITIONS NOW CONSIDERED SUITABLE FOR LANDING GROUND CREW IS READY PERIOD THUNDERSTORM OVER STATION CEILING 2000 FEET

VISIBILITY FIVE MILES TO WESTWARD SURFACE TEMPERATURE 60
SURFACE WIND WEST NORTHWEST EIGHT KNOTS GUSTS TO 20
KNOTS SURFACE PRESSURE 29.68.

At 6:22 P.M. Lakehurst radioed: RECOMMEND LANDING NOW
COMMANDING OFFICER.

At 6:44 P.M. Captain Pruss replied: COURSE LAKEHURST. . .

The *Hindenburg* dropped its lines at 7:21 P.M. Almost immediately, flames roared through the hydrogen cells, and the
frame of the ship settled to the ground. A total of 36 people died:
22 crew members, 13 passengers, and one member of the ground
crew. The cause of the disaster has never been ascertained with
finality. Many leaned to the hypothesis that some form of static
electricity ignited leaking hydrogen. Others clung to the belief
that it was an act of sabotage that sent the swastika-emblazoned
airship to its fiery ending.

Captain Pruss survived to tell the tale of the last minute:

> I was in the control car. The explosion was under the cells of gas-
> bags three and four in the aft. I heard a big noise and then I saw the
> flames. The flames were going through the whole ship and then
> forward to the bow. We were about 150 feet above the ground. My
> first idea when the explosion came was that the ropes had broken,
> but then I saw the flames and saw what it was. The stern fell and
> the bow shot up, because the gas was burning aft. I could do noth-
> ing but turn off the engines. As the control car came down to earth
> — we had under the control car a buffer, an air cushion — it
> bounced, and the ship was therefore at about a height of two meters
> [6½ feet]. The others in the crew jumped out. When the ship came
> down another time, I jumped out. Then the framework and all
> things came down, and I was under the burning cells and frame-
> work. I was in the hospital in the United States about four months.
>   It is very difficult to say what happened. The joint American-
> German Commission said: It might have been sabotage; it might
> have been lightning. Nobody knows.

The disaster to the *Hindenburg* has been compared to that
of S.S. *Titanic*. Both were ships of recent construction represent-
ing the most modern design of their types, and both were believed
by their owners to be indestructible in their realms. It turned out
differently. Though the sinking of the *Titanic* did not end the

President Wilson and officials at the start of airmail service in Philadelphia, May 1918. Smithsonian Institution Photo No. 10619.

steamer as a means of transporting people across the ocean, the tragedy befalling the *Hindenburg* marked the demise with certain finality of the airship as a commercial passenger carrier.

Flying the Mail

> Neither rain nor snow nor gloom of night stays these couriers from their appointed rounds.
> — *Herodotus**

The Army Flies the Mail in 1918

"The first plane in history to carry mail at an announced time to and from designated places on a regular schedule irrespective of weather" took to the air on May 15, 1918. The pilots were from the First Aero Squadron of the U.S. Army's Air Service, the planes were Curtiss JN6H's — the famous Jennies, and the operating agent was the United States Post Office. The purpose was to expedite the delivery of mail between Washington and New York during the existing war crisis.

---

*"The Persian messengers travel with a velocity which nothing human can equal . . . Neither snow, nor rain, nor heat, nor darkness, are permitted to obstruct their speed" (Herodotus as translated by William Beloe [1757–1817]).

The army pilots had a previous airmail experience. In supporting the Pershing Expedition into Mexico in search of Pancho Villa in 1916, they flew from Columbus, New Mexico, to the U.S. Army headquarters at Colonia Dublán in Chihuahua, a distance of some 110 miles. But even this short distance required a relay, since the Air Service at that time possessed no plane capable of staying in the air for more than one hour and twenty minutes; the cruising speed of 66 mi/h (106 km/h) gave a flying range without refueling of only 88 miles.

To carry out the new mission, existing aircraft had to be remodeled. An emergency order was placed with the Curtiss Airplane and Motor Corporation at Garden City, New York, for six JN6H's for delivery within six days. They were to be modified to carry a double-capacity gasoline tank, and the front seat was removed to make room for the mail sacks. Three of the planes were ready for delivery on May 14 and were ferried that afternoon to Philadelphia from Belmont Park Race Track on the border of New York City in Nassau County. This served as the northern terminus of the airmail service since the army's Hazelhurst Field at Mineola on Long Island was fully occupied in training military pilots for duty in France. Major Reuben Fleet, in charge of the operation, described his experience with the weather and his gas supply on the hop of 90 miles to Philadelphia:

> The weather was frightful; it was so foggy we pilots could not see each other after we left the ground. Even the masts of the boats in New York harbor were sticking up into the clouds.
>
> I climbed through the fog and came out at 11,000 feet, almost the ceiling of the plane. I flew south guided only by magnetic compass and the sun until I ran out of gas and the engine quit. Since we didn't have 'chutes in those days, there was nothing I could do but ride the Jenny down. I broke out of the clouds at 3,000 feet over lush farm land so I just picked out a nice pasture and landed. A surprised farmer sold me a five-gallon can of tractor gas but I had trouble getting it in the tank without a funnel. Perhaps three gallons got in the tank and the rest all over me, but darkness was coming and I couldn't wait while he got more from town. I asked him to point out the direction Philadelphia was and took off. Two miles from Bustleton Field I ran out of gas again and landed in a meadow. Since no telephone was available, I persuaded a farmer to drive me to Bustleton Field. Culver and Edgerton had just arrived

after similar experiences so I sent Culver with aviation gasoline to get my plane and fly it in.

Next morning Major Fleet brought a plane down to Washington from Philadelphia and landed on the Polo Field in Potomac Park soon after 10:30 A.M. President Woodrow Wilson and several dignitaries arrived to witness the historic takeoff of the first plane to fly airmail on schedule. All did not go well. The designated pilot of the plane was fresh out of flying school and secured the honored position only through his family connections with the high bureaucracy of Washington. He forgot to check his gas tank, with the result that the engine would not start. After much fussing around, someone had the wisdom to drop a measuring stick into the tank.

After a delay of forty-seven minutes before the increasingly impatient officials, Lieutenant George L. Boyle finally took off, headed in the wrong direction, became lost, and made a forced landing in a pasture some twenty miles southeast, not northeast, of Washington. The plane was damaged. It should be reported that the weather was not CAVU (ceiling and visibility unlimited), but "there was a heavy mist in the clouds which made traveling precarious," in the words of a newspaperman obviously inexperienced in reporting aviation matters. The weather map that day showed very favorable conditions for flying. A ridge of high pressure extended from James Bay in Canada to North Carolina with a cold front pushed well off the Atlantic Coast. Wind flow at Washington and New York was northerly, light to moderate.

Despite Boyle's snafu at the Washington end, the mail service was inaugurated that day by the plane coming from New York, which made the scheduled transfer of the mail sacks to a Washington-bound plane at Bustleton Field in northeast Philadelphia. The weather at the New York end was favorable for flying. The *New York Herald* reporter mentioned "the glint of the bright sun on his wings," and the *Times* man noticed a light wind of 18 mi/h (29 km/h). "The aerial express ought to be called the cloudland mail," he concluded. The temperature at takeoff time was 69°F (21°C). The elapsed time of the flight from New York to

Washington totaled three hours and twenty minutes, about two hours better than made by the steam locomotives of the day.

The experiment of having army pilots fly the mail ended at the conclusion of the ninety-day trial period on August 12, 1918. Of the scheduled flights, 92 percent had gone through. There had been fifty-five cancellations on account of bad weather, and thirty-five planes had made forced landings en route as a result of mechanical difficulties.

Private Pilots Fly the Mail

A new era for American aviation opened in August 1918 when the Post Office hired civilian pilots and assumed full responsibility for the operation of the airmail service. Many a dramatic flight was to be made as intrepid pilots fought the weather elements to get the mail through. The service soon expanded rapidly. A route was inaugurated from New York to Cleveland in August 1919 and then extended in stages to Chicago, to Omaha, and, finally, to San Francisco. All flights were by day since there were no airport floodlights or air beacons along the way. The first through flight from New York to San Francisco beat the best railroad time by twenty-two hours, despite frequent stops and the transfer of the mail sacks to the rails at night.

Insistence by Post Office officials that nothing interfere with the schedule of flights immediately led to difficulties, since they did not consider the weather a sufficient cause for not sticking to the planned departure times. The refusal of two experienced pilots to take off in foggy conditions led to their dismissal. Other pilots protested and threatened to strike. After some reluctance on the part of government officials, a conference between the two parties resulted in a compromise. It was decided to leave the decision whether to fly or not to fly to the airport manager when weather conditions were marginal.

To publicize the availability of the new coast-to-coast airmail service, the Postmaster General decided to make an experiment of transcontinental night flying with two planes leaving from San Francisco and two from New York at the same time.

The mail truck arrives with airmail at Bustleton airfield in northeast Philadelphia. Smithsonian Institution Photo No. 38903.

The date selected was Washington's Birthday, February 22, 1921. Just why the snowiest month of the year was chosen for the demonstration was never explained by officialdom, and February 1921 turned out to be a particularly stormy month.

The weather-map situation on February 22–23 was not unfavorable for such a flight, especially along the western half of the route. A large Intermountain High rested over the route from eastern California to Wyoming, and in the next twenty-four hours it strengthened to 30.56 inches (103.5 kPa). At 7:00 A.M. Chicago time, a frontal system ran from Minnesota south-southwest to western Texas, crossing the airmail route in eastern Nebraska with a band of snow. Moving eastward at about 38 mi/h (61 km/h), it reached a Buffalo-Knoxville-Mobile line by 7:00 A.M. (EST) on the twenty-third. A strong northerly flow of unstable polar air covered most of the Midwest with snow showers, and this was to cause the bad weather that one of the transcontinental planes encountered when approaching Chicago.

One of the eastbound pilots crashed and was killed in Nevada, but the other plane relayed the mail as far as North Platte, Nebraska, where Jack Knight took over and at 10:44 P.M. headed for Omaha. He ran into drizzle and a layer of clouds at 2000 feet, making it difficult to judge the horizon in the blackness or to see ahead. Since the daring flight had been announced in the press, thoughtful citizens along the route lighted bonfires to help guide him on his way. When he arrived at Omaha at 1:00 A.M., he found the entire city ablaze with lights to serve as a welcoming guide.

Knight landed safely at the airport, but the 276 miles of flying had wearied him. Upon entering the hangar, he was informed that the pilot who was supposed to fly the mail to Chicago was still weathered in at the Windy City. After some coffee and a short rest, Knight climbed into his plane again and headed east at 1:59 A.M. The ceiling measured about 1000 feet and the air became very bumpy. Snow showers reduced forward visibility to about two miles, and the horizon was again blacked out. The increasing northerly winds caused him to drift from his course.

After finding Iowa City with difficulty, Knight searched for the airport with only a road map to guide him. Again, the field was lighted by a caretaker and the pilot landed for a brief rest and refueling. When he was in the air again, the light snow turned into sleet, raising the threat of icing, the weather hazard most feared by early pilots. By flying low and following the railroad tracks that converge on Chicago, he found his way to Maywood Airport and landed at 8:40 A.M. after an ordeal of ten hours' duration.

Knight received a tumultuous welcome at Chicago and was hailed as "the ace of the air mail service." For two days the newspapers of the country carried the account of his arduous flight on the front page. The story caught the public's fancy in much the same manner that six years later another similar feat of guts and endurance by ex-airmail pilot Charles A. Lindbergh would. The mail actually arrived in New York City that afternoon. Of the elapsed time of 33 hours and 25 minutes, 25 hours and 16 minutes were consumed in flight.

Weather posed the greatest problem for the airmail pilots. Basic knowledge of aerology, the behavior of the atmospheric currents and elements at flight level, was practically nonexistent. There was no means of getting upper-air soundings of wind, temperature, and humidity and having them distributed to forecast and briefing offices to be of any real time value. Reporting of surface conditions was made at cities several times a day, but there was no frequent and systematic collection of reports from airports along the flying routes. The United States Weather Bureau, long under the control of the Department of Agriculture, made some moves to meet the new need, but not with the energy that the situation demanded. Its traditional poor relations with Congress resulted in appropriations that were woefully inadequate.

A mail pilot on the St. Louis–Chicago run, Charles "Slim" Lindbergh, was outspoken in his criticism of the forecasts and functions of the Weather Bureau. Later, in *The Spirit of St. Louis*, he employed such expressions as "paying almost no attention to weather forecasts . . . Chicago reports are so unreliable that I do not want to condition my mind with them. I would rather judge weather ahead as I fly." He claimed that weather officials did not know whether an airport was open or closed. Lindbergh had a harrowing weather experience in September 1926, when thick ground fog obscured the terrain while he was on his run to Chicago. He churned over his destination airport for thirty minutes, unable to find a hole, and then spent another thirty minutes trying to locate the edge of the fog blanket to the west and then to the south. Finally, his engine sputtered and quit. So "Slim" bailed out and parachuted safely to the ground. This was the third time in his early career that he had been forced to part company with his plane while in flight.

The difficulty of getting information about weather and field conditions ahead was illustrated by a pilot's conversation with a caretaker of a small Nebraska airfield sometime in 1923. Though the transcription, perhaps, is not verbatim, it was typical of some sources of weather reporting in those days:

> *Pilot*: "Hello, Caretaker, this is the pilot of the eastbound run to Chicago. What's your visibility there?"
> *Caretaker*: "What's my what?"
> *Pilot*: "What's your visibility? How far can you see?"
> *Caretaker*: "Ain't seein' so good this morning. Broke my glasses."
> *Pilot*: "Sorry to hear that. What's the weather like?"
> *Caretaker*: "Don't think it's so good."
> *Pilot*: "Well, what's it like?"
> *Caretaker*: "Storm's comin'. Can feel it. My corns ache like fierce."

Pilot: "That's too bad. But what kind of weather do you have right now?"

Caretaker: "Fair to middlin'."

Pilot: "Can you see across the field?"

Caretaker: "Told you. Broke my glasses."

Pilot: "Anybody else there?"

Caretaker: "Just my boy. He's fifteen. Real smart."

Pilot: "Put him on."

Boy: "Yessir?"

Pilot: "Son, look out the window of the shack and tell me how far you can see."

Boy: "'Bout thirty feet."

Pilot: "Fog?"

Boy: "No. Hay wagon parked outside the shack. Can't see over it."

Pilot (after cursing lightly to himself): "Son, do me a favor. Go outside and look up at the sky, at the clouds. See if the sun is out. Look across the field and tell me how far you can see, what object you can see that is the farthest away from you that you recognize."

Boy (after long silence): "Mister, I done like you said. Sun's out. Clouds is real purty a-rushin' past. Saw my new calf a-followin' its mother in the next pasture."

Pilot: "How far's that?"

Boy: "Can't say."

Pilot: "Why not?"

Boy: "Cow and calf is runnin' too fast."

Pilot: "Why are they running?"

Boy: "Bad storm's comin'."

Pilot: "Oh?"

Boy: "Mister, I'm goin' to have to hang up now. Pa's yellin' at me to come help him get the cows in the barn. Wind just blew the hay wagon over and our windmill's about to go. Call us back when the storm's over!"

The worst weather segment of the transcontinental run was over the Allegheny Mountains in central Pennsylvania. This became known as the "hell stretch" and "graveyard run" because of the adverse weather and the resulting accidents occurring there. The western slopes of the mountains were often saturated with moist airstreams from the Great Lakes, which cause orographic fog, rain and snow showers, and all types of icing. The eastern slopes were open to airstreams from the Atlantic Ocean bringing low clouds, drizzle, and rain when the winds blew from northeast through southeast. In the warmer season, the mountainous region itself was productive of thunderstorms, with their attendant hazards of extreme turbulence, heavy hail, and lightning. Ground fog frequently formed, especially in spring and autumn, to shroud dangerous peaks and hide valley airports. The area around Bellefonte near the crest of the mountains acquired an unsavory reputation for bad weather, and its airport proved difficult to find in poor visibility without directional radio facilities or navigation aids. It made up the center plot of the ill-reputed aviation graveyard.

The human toll of the airmail service was heavy. Thirty-four pilots and nine mechanics lost their lives while carrying the mail in the seven years of operation, most of them in the early period before the marked airways were established. It was said that the life expectancy of a pilot who joined the mail service was four years.

The year 1921 was the worst of all. Twelve experienced pilots were killed, with fifteen others seriously injured. Between April 29 and October 31, a total of eleven crackups occurred. Of the 1764 off-field landings during that year, 810 were attributed to bad weather, and the rest to mechanical difficulties. This marked the all-time high. Improvements in aircraft and development of lighted airways greatly reduced the hazardous nature of flying the mail. The odds were improving. For the first full year, 1919, one pilot died per 115,324 miles flown; during 1926, the last full year of government-operated airmail flying, there was one fatality per 2,583,056 miles flown.

The Kelly Act of 1925 marked the turning point in American aviation, which had lagged far behind European countries in the development of private, commercial, and military air activities. This legislation sought to get the government out of the airmail business and let private operators assume the sure risks and possible profits of carrying the mail. This enabled commercial airlines to get started and commence the expansion of their routes into a national network.

The Air Commerce Act, in 1926, provided for increased appropriations for government agencies concerned with fostering aviation, including the Weather Bureau. In the next year, the

Guggenheim Fund for Meteorology assisted in the expansion of weather reporting and forecasting facilities. Under the auspices of the fund, Professor Carl G. Rossby came from Bergen, Norway, where new forecasting methods showed promise of providing an objective means of making useful short-term forecasts for aviation interests. Military meteorologists from the United States had already studied in Norway, and soon even the U.S. Weather Bureau sent some of its younger men to study the new methods.

The winds of twentieth-century modernization were now stirring, wafting the Weather Bureau from its nineteenth-century agricultural focus into the aviation age that was now unfolding.

Army Pilots Fly the Mail — Again!

Early in 1934, President Franklin D. Roosevelt became irked at reports of corporate favoritism and excessive expenditures by the previous Hoover Administration in issuing contracts to commercial airlines for carrying the mails. Deciding to make some political capital out of the situation, he loosed a bombshell that had unexpected repercussions. The President ordered Postmaster General James A. Farley to cancel the contracts with the airlines and directed General Benjamin D. Foulois, commander of the Army Air Corps, to assume responsibility for flying the mail.

These actions took place on February 9, 1934 — a memorable day in meteorological history. It was the coldest morning of a severely cold month. Thermometers at many locations in the Northeast dropped to the lowest readings attained since 1857. Washington, only on the fringe of the arctic air mass, saw its official thermometer descend to $-6°F$ ($-21°C$), and was to experience its second-coldest month in history. Throughout the month, the same frigid situation prevailed over the northern half of the country from the Mississippi River to the Atlantic Ocean. On the day the flights by the army pilots were to begin, February 19, the worst blizzard of the severe winter swept the Northeast and grounded all operations.

The prevailing weather conditions over the key section of the eastern routes in Pennsylvania and Ohio were reported by the

Weather Bureau in its monthly publications for February and March 1934:

Ohio

FEBRUARY — Judged solely by State averages, the current February was the third coldest in the history of the Weather Bureau ... There were two major and two minor cold waves ... The temperature deficiency was uniformly distributed over the State ... Only four Februaries have had more snow than the current month.

MARCH — The weather throughout the month was decidedly erratic — many and quick changes from the cyclonic to the anticyclonic type, especially during the latter part of the month ... The precipitation, though deficient, was frequent and a large percentage fell in the form of snow, in fact, the snowfall was the greatest for the month of March since 1917 ... A glaze storm of decided severity covered a considerable portion of the northern half of the State, especially the northwest, from the 26th to 29th, inclusive ... In short, the winter characteristics of the month were probably sufficiently severe and numerous as to satisfy even "the oldest inhabitant" that the "old time winters" are not extinct!

Pennsylvania

FEBRUARY —... was intensely cold in all parts of the State, averaging ten degrees below normal ... At a fairly large number of stations, including Philadelphia, it was the coldest month of record ... The snowfall was unusually heavy, especially in the southeastern part of the State ... On the 19th and 20th a fairly heavy snowfall in that region was followed by gales of wind that swept the snow from the fields and piled it up behind fences and on the highways, effectually blocking traffic for two or three days ... It was a condition that has not been experienced in a generation in this part of the State.

MARCH — ... was moderately cold, and very changeable ... A jam in the Delaware, near Trenton, caused a stage of 14.2 feet on the 5th, it being the highest attained at that station since 1904. Dense fogs occurred in nearly all parts of the State during this period ... Extremes followed one another in rapid succession ... However, taken as a whole, it was a thoroughly disagreeable month, in which outdoor work made little progress.

In addition to the current meteorological difficulties, the Army Air Corps was in a state of demoralization as a result of its perennially inadequate operating budget. Most of its aircraft were of postwar, 1920s vintage. Few were equipped with instruments for blind flying or nighttime navigation, and all lacked radio facilities. Most pilots had been restricted to only four hours' flying time per month as a result of lack of funds to procure gasoline. Few had flown the transcontinental routes, with their varied terrain. So weather and inexperience were to combine to bring a string of aerial tragedies that stunned the nation.

Even before the first mail sack was carried, there were three fatal crashes, all being attributed to bad flying weather. At the end of the first week of operations, five pilots were dead, six seriously injured, and eight planes wrecked. The country was appalled at the situation. Captain Eddie Rickenbacker, the World War I ace who later headed Eastern Airlines, denounced the operation as "legalized murder," and Charles A. Lindbergh rebuked the President for his action in a stinging telegram and in subsequent congressional hearings. For the first time since the launching of the New Deal, the press and the public at large sided with the opposition; Roosevelt kept mum and let Farley take the heat.

The fatal crashes on March 9, 1934, caused the Administration to call a halt and reconsider the situation. A ten-day standdown was ordered to rest the pilots, install navigational aids, and ready some new Martin B-10 aircraft. Flights were resumed on March 19 on a curtailed scale and continued until May 8, when new temporary contracts with the airlines went into effect. With improved flying weather and the experience of the previous weeks, a much better record was attained in April and May.

Overall, the Army Air Corps experienced 57 accidents and suffered 12 fatalities while flying 1,590,155 miles and carrying a total of 777,389 pounds of mail. Altogether the army fliers completed 65 percent of the scheduled trips. Of the 1109 flights not completed, 618 were canceled because of adverse weather, and 424 were begun but not completed when conditions aloft forced pilots down. "Weather, clearly, was a villain in the piece," wrote Paul Tillot with a play on words. Few had any praise for the unfortunate Air Corps, but James Farley did point out that not a

The first weather satellite is launched at Cape Canaveral, Florida, in 1960. National Weather Service photo.

single pound of mail had been lost in the three months of flying in contrast to the airlines' previous losses of airmail pouches as a result of crashes.

The one positive good coming out of the distressing experience was the demonstration of the woeful state of the country's military air arm. Action to modernize the equipment and operations of the Army Air Corps was soon initiated. New planes were placed on the drawing boards in the midthirties that were to become the backbone of the nation's air power in World War II. "Flying the mail was a godsend for the Air Corps and the country," reminisced General Foulois later. "Without it, I am convinced we would have never recovered from the disaster of Pearl Harbor."

Space

The Space Program: Weather at the Launches

The explorations of adventurous men in the fifteenth and sixteenth centuries led to the settlement of the New World in America. This was completed in the nineteenth century with the expansion to the Pacific Coast. Now in the twentieth century Americans are engaged in the exploration of the New World of Outer Space to the bounds of our planetary system. In the twenty-first century this may result in the establishment of space stations and the colonization of the moon and planets. The first significant step by Americans in the space program was the creation of weather satellites to survey global conditions from aloft, which has led to a better understanding of our atmosphere.

The First Weather Satellite
The world's first weather satellite was launched by a Thor-Able rocket from Cape Canaveral, Florida, at 6:40 A.M., EST, on April 1, 1960. It was named TIROS, an acronym from *T*elevision and *I*nfra-*r*ed *O*bservation *S*atellite.

The morning weather placed a warm front across central Florida, running from Tampa Bay northeast to near Jacksonville.

Upper-air flow over Cape Canaveral came from the southwest at about 20 knots (37 km/h). At blastoff the total sky coverage amounted to four-tenths clouds, made up of one-tenth fog, two-tenths altocumulus at 10,000 feet, and one-tenth cirrus at an unspecified height — perfect conditions for a launch.

TIROS circled the globe once every 99.2 minutes, with its closest approach at 495 miles and farthest at 539 miles. Carrying one wide and one narrow camera, TIROS I took over 22,000 useful pictures of the earth and its cloud systems. It continued in operation until June 17, a period of 78 days.

Weather satellites of great sophistication now give hourly coverage of atmospheric and oceanic conditions of all portions of the globe.

Apollo 12 and Lightning

"The liftoff occurred in the worst weather conditions ever encountered by an American manned space vehicle," declared the *New York Times* in describing the second moon trip in Apollo 12, on November 14, 1969. Observers saw two vertical lightning-like streaks dance alongside the space vehicle shortly after liftoff at 11:22 A.M. After 36 seconds into flight a brief power shutdown occurred.

At liftoff the sky was overcast with high cirrus and a broken deck of clouds at 7000 feet and another at 11,000 feet. In addition, there were scattered clouds at 1500 feet. Very light rain showers were falling that reduced visibility to eight miles. The thermometer read 71°F (22°C) and the dew point 66°F (19°C). The wind blew from south-southwest at 12 knots (23 km/h) with gusts to 20 knots.

No lightning was indicated on the weather report for the Cape Kennedy Air Force station. Five minutes after launch the rain showers became moderate and 0.75 inch of rain fell before 12:54 A.M.

A board of inquiry convened at Houston gave the opinion that the lightning was caused by the Saturn 5 rocket itself, and not by natural causes. "We created the lightning," a spokesman declared in a press release:

The Saturn 5 rocket with the spacecraft on top was 6000 feet off the ground in a dark rain cloud when a strong lightning discharge permanently knocked out a few sensor circuits and temporarily shut off the spacecraft's computers.

"The rocket apparently triggered the lightning," Mr. McDivitt said. The clouds were not believed to have had the potential for a spontaneous discharge.

A lightning bolt occurs when the electrical charges on the earth and in a cloud differ enough to bridge the essentially nonconducting gap of air between them.

The ascending rocket carried on its surface the same charge as the ground, and its hot exhaust plume acted as a conducting wire down which the lightning traveled to the ground.

A second lightning discharge in a slightly higher cloud apparently shut down the spacecraft's guidance system.

NASA officials indicated that the agency probably would not launch a manned rocket again under similar weather conditions.

551.6973 Ludlum, David M.
LU
     The weather
     factor

DATE			